DOMESTIC DUCK PRODUCTION

DOMESTIC DUCK PRODUCTION
Science and Practice

Dr Peter Cherry

4 Chestnuts
Nettleham
Lincoln

and

Professor Trevor R. Morris

Rowan Trees
Beech Road
Tokers Green
Oxfordshire

www.cabi.org

CABI is a trading name of CAB International

CABI Head Office	CABI North American Office
Nosworthy Way	875 Massachusetts Avenue
Wallingford	7th Floor
Oxfordshire OX10 8DE	Cambridge, MA 02139
UK	USA
Tel: +44 (0)1491 832111	Tel: +1 617 395 4056
Fax: +44 (0)1491 833508	Fax: +1 617 354 6875
Email: cabi@cabi.org	Email: cabi-nao@cabi.org
Web site: www.cabi.org	

A catalogue record for this book is available from the British Library, London, UK.

Library of Congress Cataloging-in-Publication Data

Cherry, P. (Peter), 1937-
  Domestic duck production : science and practice / P. Cherry and T.R. Morris.
    p. cm.
  Includes bibliographical references and index.
  ISBN 978-0-85199-054-5 (alk.paper)
1. Ducks. I. Morris, T.R. (Trevor Raymond). II. Title.
SF505.C55 2008
636.5'97—dc22

                                                                                    2008006764

ISBN-13: 978-0-85199-054-5 (HB)

ISBN-13: 978-1-84593-955-7 (PB)

First published (HB) 2008
First paperback edition 2011

Printed and bound in the UK by MPG Books Group.

Contents

Preface

This book is intended as a practical guide for technical and advisory staff involved in both commercial and subsistence-level duck production, as a reference text for commercial producers and their professional staff and as a resource for undergraduate and postgraduate students.

The text describes research conducted and experience gained in differing climates and economic conditions between 1975 and 2005. Genetic selection and technical innovation have dramatically improved duck performance during those 30 years, but that has not substantially altered the concepts and relationships reviewed and described in this book. The science remains the same, although practice evolves.

Acknowledgements

The authors are grateful to Cherry Valley Farms Ltd, Maple Leaf Farms, Inc and other commercial companies in the USA, Europe and South-east Asia that provided the resources for the numerous trials described. We also wish to thank the many colleagues and staff who gave generous assistance down the years.

We are grateful to Peter Lavery for kindly providing the photograph used on the front cover and to Geoff Espin for his assistance in preparing black and white photographs from original colour transparencies.

We thank the Director, staff and Librarian at the Northern Prairie Wildlife Research Centre, North Dakota, USA, and the Director and Dr Cynthia Bluhm of the Delta Waterfowl and Wetlands Research Station, Portage la Prairie, Manitoba, Canada, for their generous help with our many enquiries about wildfowl. We also wish to thank the Director and staff of the Veterinary Agrotechnology Department, Singapore, for their cooperation and assistance in carrying out an experiment at that centre and Professor Rob Gous of the University of Kwazulu-Natal, Pietermaritzburg, South Africa, for providing nutritional analyses.

Dr P. Cherry would like to thank his wife Audrey for her constant support and assistance, and for accompanying him during a study of wildfowl in wetlands and wilderness areas of the USA, Canada and Australasia. Thanks also to Brenda Curley and to my daughters Sarah and Helen and grandchildren Ellie, Tasha and Archie whose unfailing good humour has been of inestimable value throughout this study.

Finally, I would like to thank my father W.J. Cherry who taught me from an early age the craft of poultry husbandry and Professor Trevor Morris whose help, advice and encouragement over many years made this book possible.

Lincoln
2008

1 History and Biology of the Domestic Duck

When the first settlers arrived in Australia in the 18th century, the aboriginal people lived as hunter-gatherers. They collected duck eggs during the breeding season and trapped adult ducks when they moulted and became temporarily flightless following the summer solstice. At other times of the year ducks were hunted on the wing with the aid of boomerangs. Birds on water were captured with slip nooses hidden among reed beds and caught by hunters swimming underwater, breathing with the aid of reed tubes (Flood, 1996). Australian aboriginals hunted duck in this manner over many thousands of years and it seems very probable that hunter-gatherers in other parts of the world employed similar methods.

Paintings and relief carvings in the tombs of Mereruke, vizier to Pharaoh Teti (6th dynasty, 2240 BC) at Saqqara, and Nacht (18th dynasty, 1550 BC), scribe and 'Astronomer to Amun' at the Karnak Temple in Egypt, show that more than 3000 years ago migratory wildfowl were hunted and trapped with large, hexagonal-shaped clap-nets in the extensive swamplands of the Nile delta. These ducks were kept in large aviaries and were force-fed before slaughter to provide a ready supply of meat throughout the year. However, although there is some evidence to suggest that native geese were domesticated in Egypt, it seems most unlikely that ducks were domesticated until much later.

In the century before Christ, Marcus Terentius Varro (116–27 BC), a Roman who wrote extensively on agricultural subjects and the earliest writer to describe raising ducks, suggested 'it was necessary to build a solid enclosure covered with a wide-meshed net so that an eagle could not fly in or the ducks fly out'. Lucius Moderatus Columella, writing at about the same time, suggested 'anyone wishing to establish a place for rearing ducks should collect wildfowl eggs in the marshes and set them under farmyard hens, for when they are reared in this way they lay aside their wild nature and without hesitation breed shut up in the bird pen'. He suggested '"amphibious" birds should be fenced in, provided with swimming water and a range area clothed with grass and supplied with nest boxes made of stone covered with a smooth layer of plaster in which birds may lay their eggs'. Marcus Porcius Cato also provided similar advice but suggested 'ducks should be fed wheat, barley, grape marc and sometimes even lobsters and other aquatic animals, and the pond in the enclosure should be fed with a large head of water so that it may always be kept fresh'.

Paintings, mosaics and other archaeological evidence, including duck bones excavated from middens at many locations in the Roman Empire, indicate that duck was eaten, probably on feast days and festivals. However, it seems likely that ducks, unlike chicken and geese, were not domesticated, but simply held captive along with peafowl, guinea fowl and fieldfares, to provide occasional exotic food for Roman tables. Both feather and down were used as insulation in clothing and bedding. Primary feathers, commonly termed 'quills', were used to stabilize the flight of arrows essential for hunting, and sharpened goose quills were used for writing.

Harper (1972) reviewed domestication of duck in the classical and medieval literature, but was unable, apart from the texts previously described by Romans, to find any clear reference to domestic, as opposed to wild, duck until about AD 810, when they begin to appear listed along with geese, chicken and peafowl as produce-rents, paid by peasants to landowners of large estates. Descriptions of domestic farm animals in manuscripts dated around 1150 include both domesticated geese and ducks, but ducks are not mentioned in any of the medieval treatises on farming, supporting the opinion of Delacour (1956) that ducks were not domesticated in Western Europe much before the Middle Ages.

Domestic ducks are increasingly mentioned in the financial records of large estates in England from about the 12th century, but it seems they were of little financial consequence. Their value may have been affected by plentiful supplies of wild duck because chronicles written in the 13th century confirm that wildfowl were caught with nets in the fenlands of Eastern England when they became flightless during their eclipse moult following the summer solstice. The quantities caught were sufficient to provoke legislation in 1534, imposing a closed season (Robinson, 2004).

By about the 15th century, decoys were increasingly used to assist in trapping wildfowl migrating between England, the Baltic and Russia. Tame and vociferous 'Call' decoy ducks were placed on ponds or tidal creeks leading into curved tunnel nets. When sufficient wild ducks were attracted by the decoys, a decoy man hidden behind screens would encourage a small dog of foxy appearance to leap backwards and forwards along the sides of the tunnel net. This behaviour aroused the curiosity of wildfowl, and driven by their 'mobbing' instincts they would pursue the dog into and along the curved tunnel net (about 50–60 m in length) where they were caught, killed and then transported to market.

Robinson (2004) confirms there were as many as 215 'decoys' located along the eastern counties of England with several tunnel nets radiating from a central trap. Decoys provided an estimated 'catch' of more than a million ducks per year and continued to be used until towards the middle of the 19th century. In addition, punt guns capable of killing large numbers of ducks with a single shot were introduced during the 16th century, substantially increasing the numbers of wildfowl killed and sold in markets across England.

Pictures of domestic duck appear in the 17th century and a painting in the National Gallery in London by Jon Steen (1660) entitled 'The Poultry-Yard' includes white-feathered domestic ducks. Domesticated ducks appear increasingly in landscape paintings throughout Western Europe, indicating they were probably widely distributed throughout the area by the end of the 17th century.

The population of the major cities in England grew substantially over the 18th and 19th centuries, increasing demand for all types of poultry and game. Production of ducks became centred in particular areas, the most famous in England being Aylesbury (see Chapter 2, this volume), where selective breeding produced the well-known Aylesbury domestic duck.

Domestication in China and the Far East

Wucheng (1988) suggests pottery ducks excavated in the Yan-shi-menkou Mountains in Fu Jian Province in southern China provide evidence that domestication of duck occurred during the New Stone Age between 4000 and 10,000 years ago. Jianhua (2004) describes clay models unearthed in Shanxian county, central Henan Province, and suggests ducks were domesticated during the Yangshao culture about 4000 years ago. Clayton (1984) describes a report in the Chinese literature by Yeh, who investigated the archaeological evidence and suggested ducks were domesticated in China at least 3000 years ago.

Whilst there appears to be little written evidence regarding the history of domestication, Jung and Zhou (1980) confirm that, to satisfy the preference of high officials and feudal landlords for roast duck, farmer's carried out careful selection. During the Ming dynasty (AD 1368–1644) the Pekin duck was already known as an outstanding breed with stable genetic characters.

Peng (1984) describes a report by Tao Huo (1487–1540) that 'domestic ducks were used to excellently control amphibious crabs in rice paddies' and reviews descriptions provided by Jinglun Chen in 'Locust Prevention Notes' written in 1597 and by Shiyi Lu (1611–1672) in 'Locust Killing Notes', confirming domestic ducks were used to control locusts, leafhopper, rice shield bug and many other pests in Zhujiang River rice paddies.

Clayton (1984) suggests the history of domestication of the common duck in both China and Western Europe is obscure but, like the chicken, the range of types emanating from the Far East suggests South-east Asia as a major centre of domestication. The archaeological evidence along with a favourable environment and agriculture suggest that ducks were probably domesticated in southern China at least 1500 years before they were separately domesticated in Western Europe.

Mallard: Progenitor to Domestic Duck

Both Darwin (1883) and Delacour (1956) consider that the domestic duck derives from the characteristically green-headed wild Mallard, *Anas platyrhynchos*, which is widely distributed over the northern hemisphere. Mallard are included among dabbling ducks or surface feeders in the tribe Anatina, subfamily Anatinae of the Anatidae, which includes most wildfowl.

Darwin was interested in both the origin and variation in breeds of domestic duck and described the downward curving bill of the hook-billed duck, first described in 1676 and originating in China as extraordinary. He also described the physical appearance of domestic breeds such as the Aylesbury, Rouen, Call, Polled and Labrador duck, but suggested the Penguin duck (Indian Runner) was the most remarkable of all breeds because of the way this small, thin bird walked with its body extremely erect and its thin neck stretched upwards.

Darwin described how wild Mallard were easy to tame, crossed readily with domestic breeds and produced fertile offspring. He confirmed that body-weight along with the width of the white feather neck collar increased rapidly when bred in captivity and birds soon lost the ability to fly. He confirmed that four middle tail feathers curl upwards in the drakes of only one wild species, *A. platyrhynchos* (wild Mallard), and that this characteristic feathering can be seen in the tail feathers of drakes of *all* domestic breeds. He suggested there appeared little doubt that the wild Mallard was the parent of all breeds of domestic duck.

The Natural History of Wild Mallard

The brilliant green glossy head and neck, white collar, rich chestnut-coloured chest feathers, white and brilliant blue speculum in the wing and bright orange bill, legs and feet of the drake along with their wide distribution across the northern hemisphere means Mallard drakes (live weight about 1250 g) are easily recognized and the best known of all wildfowl. Female Mallard (live weight about 1100 g), by contrast, are drab brown with orange legs and feet, but are easily recognized by their characteristic white and brilliant blue speculum which, beside a display function, probably assists visual contact during migration, because most migratory wildfowl have white feathers located amongst either their wing or tail feathers (Ogilvie and Pearson, 1994).

Mallard obtain much of their feed by sieving particles of food from surface water. Moving the tongue rapidly up and down encourages a pumping action sucking in a mixture of food and water through the open tip of their bill and then squirting the mixture through fine comb-like structures (lamellae) located evenly about 1 mm apart along the sides of the bill. This traps seeds and invertebrates. Birds can sometimes be seen paddling vigorously in shallow water to encourage vegetable matter and insects to the surface; this increases the supply of available feed and also allows birds to feed in almost complete darkness.

Dabbling ducks are gregarious. Feeding together in flocks provides some safety from predators, particularly when searching for invertebrates below the surface of the water, or when 'up ending' and searching for weed and crustaceans in the silt along the margins of lakes and pools. However, all wildfowl peck and are able to pick up individual items of food. Mallard can swallow slugs and snails of more than about 1 cm diameter. They are opportunistic feeders, eating a wide range of aquatic vegetation, invertebrates and insects, and glean stubble fields for grain following harvest in autumn.

Dabbling ducks can take off from water or land, and Queeny (1983) in a comprehensive study illustrates how Mallard, when threatened, spread their wings and use the water surface to spring vertically into the air, literally flying out of the water to achieve a rapid and noisy exit. Ducks tolerate cold and icy water by restricting blood flow to their legs and feet, and in hot weather increase circulation and heat loss from legs and bill. Mallard are hardy but prefer ice-free water and migrate south before the onset of winter to coastal estuaries and ice-free inland waterways. Pairing takes place towards the end of the year, as birds migrate north and return to the breeding grounds in early spring, often to the locality where females were hatched and reared.

As a consequence of the relative importance of wildfowl in the USA and Canada there is an extensive literature on factors affecting sexual maturity and reproduction of wild Mallard (Bluhm, 1992; Nielson, 1992). Timing of the breeding season in wildfowl is primarily controlled by photoperiod (Murton and Westwood, 1977). Batt and Prince (1978a) found that initiation of first nest for captive females was at the same time of the year over three separate years. However, the effect of photoperiod is moderated by environmental factors, primarily access to food supplies (Perrins, 1970; Krapu, 1974, 1979, 1981; Pehrsson, 1984; Johnson and Grier, 1988; Mitchell, 1992).

Birds arrive at the breeding grounds following migration carrying substantial reserves of body fat (Krapu, 1981), but are dependent upon a ready supply of protein for egg formation. Swanson et al. (1979) found that females select animal foods to satisfy their demand for protein for egg production in contrast to males, who subsist mainly on a vegetable diet.

Age has also been shown to affect date of nesting, yearling females breeding several weeks later than 2-year-old and older females (Batt and Prince, 1978b). A significant effect of temperature upon sexual maturity has been reported by Hill (1984), Krapu (1979) and Langford and Driver (1979), who concluded that a 1° decrease in average minimum temperature in early spring was associated with a delay of about 1.2 days in time of peak nest initiation. However, it seems likely that the principal effect of temperature is through its effect upon the availability of suitable food supplies for egg formation. Pehrsson (1984) investigated the responses of Mallard to climate change in Sweden and concluded that sub-aquatic invertebrates were essential for successful egg formation, and emerging insects along with seeds and vegetative food provided nutrients for hatchlings and growing birds.

Female bodyweight declines by as much as 20% through the nesting season (Krapu, 1981) associated with the energy demands of egg production and reduced feeding time as a consequence of incubation. Clutch size is largely affected by the available food supply. The average clutch size in North America is 10–11 eggs with an average egg weight of about 50 g (Krapu and Doty, 1979; Swanson et al., 1986) with females capable of four renests when eggs are lost to predation or farming cultivations (Lokemoen et al., 1990a). Wild Mallard held in captivity can lay as many as 90 eggs per female prior to moulting in early summer (Hunter and Scholes, 1954).

Incubation takes about 28 days, commencing after the last egg is laid to enable all chicks to hatch more or less together. Drakes desert their mates shortly after the start of incubation, relocating to marshland or open water. Here they lose their breeding plumage and become temporarily flightless during the 'eclipse moult', when they lose all their primary and secondary flight feathers. They seek safety among the reeds and shrubs close to open water until their flight feathers have regrown.

As soon as hatching is complete females lead their brood (average chick weight about 30–34 g) to water, but make no attempt to feed them. Chicks forage for small insects and plants such as duckweed, and can be observed 'up-ending' as early as 7 days of age. Brooding females shelter chicks under their wings at night and during bad weather, until such time as they are no longer able to cover them (Ogilvie and Pearson, 1994). Young Mallard grow rapidly (see Fig. 5.37), fledging by about 50 days and achieving adult mature live weight at about 12–14 weeks of age, by which time feathering with maturation of the primary and secondary wing feathers is complete, enabling them to undertake the annual migration to overwintering areas.

Females abandon their broods shortly after they fledge and then moult in time to complete the annual migration in the autumn. The rapidly decreasing daylength encourages young and old to gather on open water and together they migrate south in large flocks seeking ice-free open water.

Genetic Inheritance from Mallard

Chapter 5 (this volume) describes and compares the relative growth and composition of Mallard and Pekin and shows that, despite domestication and selection over perhaps 2000 or 3000 years, domestic ducks still have many similarities with their wild forebear, the Mallard. Figures 5.37–5.40 show that despite substantial differences in absolute rate of growth of bone, body weight and composition, and flight feathers, the pattern of growth and time to somatic maturity remain remarkably similar for wild and domesticated ducks.

In Europe and North America, wild Mallard breed the year after hatching, when they are about 50 weeks old. Domestic duck reared on either *ad libitum* or restricted feed lay their first egg by about 17 weeks. Cherry (1993) reared captive Mallard and two Pekin genotypes selected for egg production and growth, respectively, on *ad libitum* feed and showed that, when given a 23 h daylength from day-old to 8 weeks, which was then reduced weekly in equal steps to provide 17 h at 18 weeks, Mallard and Pekin responded in a similar manner laying their first eggs at about 21 and 17 weeks, respectively. These results, along with evidence presented in Chapter 6 (this volume), confirm that daylength during rearing, and not age, is the principal factor controlling age at sexual maturity for both Mallard and domesticated ducks.

Mallard breed early in the year to enable their young to mature sufficiently to migrate in the autumn. The increasing daylength in early spring promotes sexual maturity and breeding activity, but towards the summer solstice also encourages both sexes to become photorefractory, preventing breeding pairs from producing progeny insufficiently mature to migrate in the autumn. However, as a result of centuries of selection for laying perform-ance, domestic breeds given either an increasing natural daylength or step-up lighting during rearing no longer become fully photorefractive. Nevertheless, evidence presented in Chapter 6 (this volume) shows that increasing daylength during the latter stages of rearing still affects the laying performance of domestic duck, substantially reducing their post-peak rate of lay (see Figs 6.36 and 6.39).

Geographic Distribution

Total duck production from countries reported in the Food and Agriculture annual published statistics for 2004 (FAO, 2004), which does not include the Russian Federation, Ukraine and other former Soviet countries, is about 2249 million ducks per annum. However, this total includes Muscovy (*Cairina mos-chata*), a native duck of South America, originally domesticated by the Colombian and Peruvian Indians and introduced to Europe by the Spanish and Portuguese in the 16th century, and sterile hybrids of Muscovy and Pekin (mules) used for liver production in France and some countries in the Far East.

China produces more than 1.7 billion domestic ducks per annum, about 73% of the world production. China along with the South-east Asian coun-tries of Vietnam (3.3%), Thailand (2.5%), Taiwan (1.5%), Indonesia (1.2%), Republic of Korea (1%), Myanmar (0.9%) and the Philippines (0.4%) account for about 85% of world production. India, Pakistan and Bangladesh produce about 2.8%; France (3.6%) along with countries of the European Community (EC) produce about 6.5%; and North America about 1.7% of recorded world supplies.

Table ducklings are popular in South-east Asia and most visitors to China will have enjoyed the world-famous Beijing 'Pekin' roast duck. While there are no published statistics for the production of duck eggs, they are widely eaten and used in South-east Asian cuisine. Duck eggs are also further processed. For example, 'Century' eggs, made by placing fresh duck eggs in a saline pickling solution (NaCl and NaOH) at about 15°C for about 20 days before coating them with red mud and then aging them for several weeks, are a popu-lar and traditional Chinese delicacy. In Vietnam and the Philippines, 'balut', fertilized eggs incubated for about 18 days and then lightly boiled in water, are widely enjoyed.

Duck Breeds

Ducks have been used for centuries to forage for insect pests over rice paddy and this no doubt encouraged the widespread practice of 'droving' in many parts of South-east Asia. Flocking behaviour, inherited from their wild forebears, allows ducks to be herded. Birds forage in paddy fields on their way to market, harrowing and cleaning land prior to planting, scavenging growing crops for pests and insects and gleaning after the harvest, a journey that may cover many miles and take several weeks, and probably accounts for the erect posture and running ability of breeds such as the Indian Runner from Malaysia and Bali from Indonesia along with the mobility and stamina of breeds of native ducks such as the Shaoxing, Jinding and Maya from China.

Many indigenous breeds are robust, capable of high levels of egg production (see Table 7.1) and enjoy exceptional down quality (measured in terms of both down cluster size and insulation Tog value) and weatherproof feathering. All have their own local names and, although often relatively unknown, enjoy substantial reputations in South-east Asia and deserve to be better known in the rest of the world.

Clayton (1984), reviewing a report by Kingston *et al.* (1979) on the management of laying duck flocks in the swamps of South Kalimantan, Indonesia (where the local native breed of Alabio achieved 245 eggs in 365 days), commented that 'the exceptional qualities shown by native breeds are entirely attributable to their history in the hands of the peasants of South-east Asia and owe nothing to Western genetics and technology'.

The Khaki Campbell is perhaps the best known laying breed, named after Mrs Campbell who, towards the end of the 19th century, acquired a single Indian Runner duck that was reputed to have laid many eggs and crossed it with a Rouen, a large meat-type French breed. She crossed a subsequent generation with wild Mallard and after several generations extracted an intermediate-sized khaki-coloured duck (Clayton, 1984). The breed soon acquired a reputation for egg laying. In 1923 Jansen, a Dutch breeder, recorded 240 eggs in a laying year in a flock of recently imported Khaki Campbell ducks (Clayton, 1972). Using genetic selection, Jansen substantially increased laying performance, with flocks producing as many as 335 eggs in 52 weeks of lay (Hutt, 1952).

Nho and Tieu (1996) reported trials in Vietnam in which flocks of Khaki Campbell ducks were either confined in pens on ponds or canals and fed paddy rice with seasonally available seafood, or were herded along the tidal seashore on the Red River delta and provided with rice grain during the days of ebb, but confined during flow and given shrimp and fish. Confined and herded flocks laid their first egg at 156 and 144 days, and achieved 242 and 271 eggs in 52 weeks with an average egg weight of 70 and 71 g and consumed an average of 2.25 and 1.87 kg of paddy rice and 1.87 and 0.69 kg of sea feeds for every 10 eggs produced, respectively. Thus, both confined and herded Khaki Campbell ducks provided with local feedstuffs achieved excellent performance, which confirms that the Khaki Campbell is hardy, tolerates high temperatures and is capable of foraging.

In Taiwan, the Tsaiya, originally domesticated in China and brought to the island by early settlers, occurs in white, coloured and Mallard-type feathering and can produce more than 250 eggs per annum (Huang, 1973). There are many strains of this famous breed, and both coloured and white-feathered strains are crossed with Muscovy drakes (introduced to the island during the 18th century) with a mature bodyweight more than 4.5 kg. Both natural mating and artificial insemination are used to produce sterile Mule ducks which are raised mainly for meat.

The Pekin is the best known of all table breeds and originated in China as early as the Ming dynasty (1364–1644) (Jung and Zhou, 1980). It was introduced to the USA in about 1873 (Scott and Dean, 1991) and to Europe a few years later, and has since become the predominant table breed in many parts of the world. In China, production is sometimes integrated with fish farming (see Chapter 2, this volume) but, in major cities, table ducklings are increasingly reared intensively to meet the requirements of an expanding and aspiring population. The effect of selection on performance is reviewed in the following section of this chapter.

The historic evidence suggests that ducks were domesticated independently in Europe, but much later than in China, and they have never achieved the same economic importance in the Western world as in South-east Asia. Consequently, there are fewer European breeds of duck. The Aylesbury duck appears to have originated in the 18th century and was for many years the most popular English breed and enjoyed a considerable reputation as a table bird until displaced by the Pekin after 1945. Other European breeds include the diminutive 'Call' duck (mature bodyweight less than 1 kg) with a loud and penetrating voice, used since medieval times as a decoy, the Rouen, a large and brown-feathered French breed, and the world-famous Khaki Campbell.

Domestic breeds occur in a wide range of colours, but pigmented 'pin-feathers' detract from the appearance and the subsequent market value of processed birds. White feathers and down are used for bedding and clothing and command a premium price. As a result, breeds of commercial importance in North America and the EC are all white-feathered. The Call, Pekin and Aylesbury duck breeds are homozygous for recessive white, and Clayton (1984) reports that if any of these breeds is crossed with a coloured breed, the first generation will be coloured. If mated *inter se*, these first crosses produce about 25% white-feathered progeny. Backcrossing these white segregants to the coloured grandparental breed and repeating the process a number of times produces a genotype virtually identical with the original coloured strain, but solid white in feather colour. This process may be the explanation for white-feathered varieties of well-known coloured breeds, for example White Campbell, a derivative of the world-famous Khaki Campbell used to provide laying performance in modern female hybrid genotypes.

Recent History and Performance

In Asia, duck production integrated with both rice and fish farming has continued in much the same way for centuries, providing an important source of food and income for farming communities (FAO, 2007). Historically, much of the research and reviews on all aspects of duck production have been published in Chinese, limiting access by farmers and scientists located overseas. However, rapid economic development in China and the Far East, along with better access, transport and communication via the Internet, allows farmers in China,

South-east Asia and Western economies to exchange information, compare experience and improve their economic performance.

In Taiwan and France, both major duck-producing countries, demand for meat quality and duck liver has encouraged research, development and production of Muscovy × Tsaiya and Muscovy × Pekin hybrids, known as 'Mule' and 'Moulard', respectively. In Taiwan, the meat from Mule ducks enjoys an excellent reputation, providing abundant breast meat of excellent quality and flavour. Moulard provide breast meat fillets and allow French restaurants to produce the world-famous 'Magret of duck'. In a designated area of France, force-feeding provides duck liver of sufficient size and quality to satisfy the demands of both domestic and international markets.

In the USA, Lee (1915) provided the first record of performance and guidance on husbandry. Birds were reared with access to grass and given supplementary feed. They were slaughtered between 8 and 12 weeks at weights of 2–2.7 kg and were referred to as 'green ducks'.

The first detailed record of growing performance for Pekin duck was published by Horton (1928). Figure 1.1 shows a large improvement in weight for age over time for genotypes from the USA and EC, and Fig. 1.2 shows that live weight at 7 weeks increased almost linearly between 1928 and 2001. Figure 1.3 describes the relationship between weight and feed intake. Selection for growth, along with improved nutrition, reduced age at slaughter at a weight of about 2.7 kg by roughly 40 days between 1928 and 1993 and there was a large correlated improvement in the efficiency of feed conversion (Fig. 1.4).

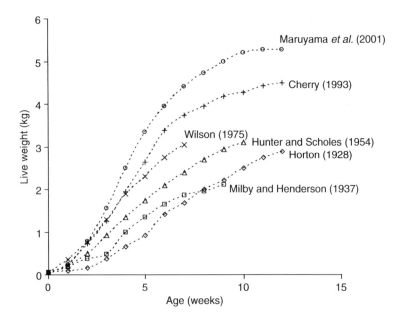

Fig. 1.1. Effect of age on growth of Pekin recorded in the USA and EC between 1928 and 2001.

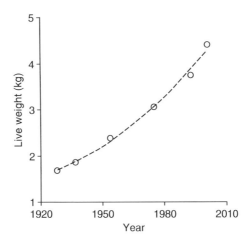

Fig. 1.2. Live weight for Pekin recorded at 7 weeks of age. (Data sources as in Fig. 1.1.) Live weight increased almost linearly by about 36 g per year between 1928 and 2001.

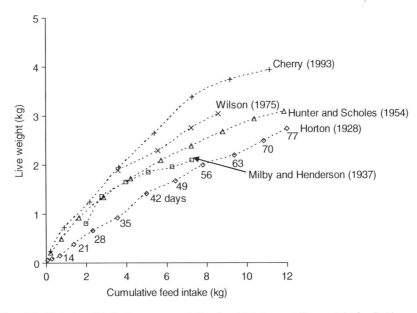

Fig. 1.3. Relationship between cumulative feed intake and live weight for Pekin recorded in the USA and EC between 1928 and 1993.

Selection for increased weight at slaughter in Pekin ducks has had the effect of increasing potential mature adult weight, but the majority of breeding stock within the USA and EC are now reared upon some form of feed restriction to avoid precocious development and to synchronize sexual maturity, limit adult size and improve reproductive fitness.

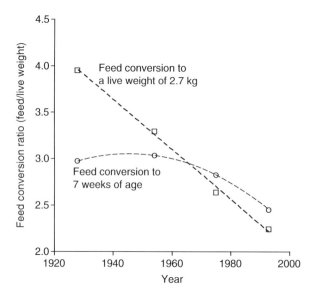

Fig. 1.4. Feed conversion ratio for Pekin reared to 7 weeks of age and to a fixed live weight of 2.7 kg. Age to achieve a live weight of 2.7 kg declined linearly from 75 days in 1928 to about 35 days by 1993, a reduction of about 0.6 days per annum, and reduced the amount of feed to produce 1 kg of bodyweight by about 27 g per annum over the period 1928–1993. (Data as described in Fig. 1.3.)

References

Batt, B.D.J. and Prince, H.H. (1978a) Laying dates, clutch size and egg weight of captive Mallard. *Condor* 81, 35–41.

Batt, B.D.J. and Prince, H.H. (1978b) Some reproductive parameters of Mallards in relation to age, captivity and geographic origin. *Journal of Wildlife Management* 42, 834–842.

Bluhm, C.K. (1992) Environmental and endocrine control of water fowl reproduction. In: Batt, B.D.J. (ed.) *Ecology and Management of Breeding Waterfowl*. University of Minnesota Press, Minneapolis, Minnesota, pp. 323–364.

Cherry, P. (1993) Sexual maturity in the domestic duck. PhD thesis. The University of Reading, Reading, UK.

Clayton, G.A. (1972) Effects of selection on reproduction of avian species. *Journal of Reproduction and Fertility* (Suppl 15) pp. 1–21.

Clayton, G.A. (1984) Common duck. In: Mason, I.L. (ed.) *Evolution of Domesticated Animals*. Longman, London, pp. 334–339.

Darwin, C. (1883) *The Variation of Animals and Plants under Domestication*, 2nd edn. Appleton, New York.

Delacour, J. (1956) *The Waterfowl of the World*. Vols 2–4. Country Life, London.

FAO (2004) Global Livestock Production and Health Atlas – GLiPHA. Available at: www.fao. org/ag/aga/glipha/index.jsp?page=home.html

FAO (2007) The role of scavenging ducks, duckweed and fish in integrated farming system in Vietnam. Available at: www.fao.org/ag/againfo/resources/document/frg/conf96htm/ men.htm

Flood, J.M. (1996) *The Moth Hunters of the Australian Capital Territory*. J.M. Flood, Canberra, Australia.

Harper, J. (1972) The tardy domestication of the duck. *Agricultural History* 46 (3), 385–389.

Hill, D.A. (1984) Laying date, clutch size and egg size of the Mallard *Anas platyrhynchos* and Tufted duck *Ayatha fuligula*. *Ibis* 126, 484–495.

Horton, D.H. (1928) The growth of white Pekin ducklings. *Poultry Science* 7, 163–167.

Huang, H. (1973) The duck industry of Taiwan. *Animal Industry Series No.8*, Chinese-American Joint Commission on Rural Construction, Taipei, Taiwan.

Hunter, J.M. and Scholes, J.C. (1954) *Profitable Duck Management*, 9th edn. The Beacon Milling Company, Cayuga, New York.

Hutt, F.B. (1952) The Jansen Khaki Campbell ducks. *Journal of Heredity* 43, 277–281.

Jianhua, W. (2004) Tri-colored duck potteries of the Tang dynasty. *China and World Cultural Exchange* 69, 1. Available at: http://www.zwwhjl.com.cn

Johnson, D.H. and Grier, J.W. (1988) Determinants of breeding distributions of ducks. Wildlife Monograph 100, supplement to *Journal of Wildlife Management* 52, 1–37.

Jung, Y. and Zhou, Y.P. (1980) The Pekin duck in China. *World Animal Review* 34, 11–14.

Kingston, D.J., Kosasih, D. and Ardi, I. (1979) The rearing of Alabio ducklings and management of laying duck flocks in the swamps of South Kalimantan. *Report No.9*. Centre for Animal Research and Development, Ciawa, Bogor, Indonesia.

Krapu, G.L. (1974) Feeding ecology of Pintail hens during reproduction. *Auk* 91, 278–290.

Krapu, G.L. (1979) Nutrition of female dabbling ducks during reproduction. In: Bookhout, T.A. (ed.) *Waterfowl and Wetlands – An Integrated Review*. North-central Section, The Wildlife Society, Madison, Wisconsin, pp. 59–70.

Krapu, G.L. (1981) The role of nutrient reserves in Mallard reproduction. *Auk* 98, 29–39.

Krapu, G.L. and Doty, H. (1979) Age related aspects of Mallard reproduction. *Wildfowl* 30, 35–39.

Langford, W.A. and Driver, E.A. (1979) Quantification of the relationship between nest initiation and temperature. *Wildfowl* 30, 32–34.

Lee, A.R. (1915) *Duck Raising*, United States Department of Agriculture Farmers Bulletin, No 697, Government Printing Office, Washington, DC.

Lokemoen, J.T., Duebbert, H.F. and Sharpe, D.E. (1990a) Homing and reproductive habits of Mallards, Gadwalls and Blue-Winged Teal. Wildlife Monograph 106, supplement to *Journal of Wildlife Management* 54, 1–28.

Lokemoen, J.T., Johnson, D.T. and Sharp, D.E. (1990b). Weights of wild Mallard, Gadwall and Blue-winged Teal during the breeding season. *Wildfowl* 41, 122–130.

Maruyama, K., Vineyard, B., Akbar, K., Shafer, D.J. and Turk, C.M. (2001) Growth curve analyses in selected duck lines. *British Poultry Science* 42, 574–582.

Milby, T.T. and Henderson, E.W. (1937) The comparative growth rates of Turkeys, Ducks, Geese and Pheasants. *Poultry Science* 16, 155–165.

Mitchell, J.T. (1992) Our disappearing wetlands. *National Geographic Magazine* 182, 3–45.

Murton, R.K. and Westwood, N.J. (1977) *Avian Breeding Cycles*. Clarendon Press, London.

Nho, L.T and Tieu, H.V. (1996) Egg production and economic efficiency of Khaki Campbell ducks reared on locally available feedstuffs in the coastal land stretch of the Red River Delta. *Livestock Research for Rural Development* 9 (1), 1–4.

Nielson, S. (1992) *Mallards*, Swann Hill Press, Shrewsbury, UK.

Ogilvie, M. and Pearson, B. (1994) *Wildfowl*, Hamlyn, London.

Pehrsson, O. (1984) Relationships of food to spatial and temporal strategies of Mallard in Sweden. *Journal of Wildfowl Management* 48, 323–339.

Peng, S. (1984) Pest control methods in ancient Chinese agriculture. *Agricultural Archaeology* 2, 266–268.

Perrins, C.M. (1970) The timing of birds breeding seasons. *Ibis* 112, 242–255.

Queeny, E.M. (1983) *Prairie Wings. The Classic Illustrated Study of American Wildfowl in Flight.* Dover Publications, New York.

Robinson, K. R. (2004) Duck decoys in Kent. *Kent Archaeological Review* 156, 125–128.

Scott, M.L. and Dean, W.F. (1991) *Nutrition and Management of Ducks.* M.L. Scott of Ithaca, Ithaca, New York.

Swanson, G.A., Krapu, G.A. and Serie, J.R. (1979) Foods of laying female dabbling ducks on the breeding grounds. In: Bookhaut, T.A. (ed.) *Waterfowl and Wetlands – An Integrated Review.* North-central Section, The Wildlife Society, Madison, Wisconsin, pp. 47–57.

Swanson, G.A., Shaffer, T.L., Wolf, J.F. and Lee, F.B. (1986) Renesting characteristics of captive Mallards on experimental ponds. *Journal of Wildlife Management* 50, 32–38.

Wilson, B.J. (1975) The performance of male ducklings given diets with different concentrations of energy and protein. *British Poultry Science* 16, 617–625.

Wucheng, B. (1988) The research on the origin of the house-duck in China. In: *Proceedings of the International Symposium on Waterfowl Production.* The Satellite Conference for the 18th World's Poultry Congress, Beijing, China. Pergamon Press, Oxford, pp. 125–129.

2 Systems of Production

Until the 20th century, ducks were kept in small flocks by smallholders, farmers or estate owners, mainly as a minor source of food for the household. Some 18th- and 19th-century European gentlemen maintained ducks, and imported many foreign varieties for their ornamental value, but nowhere in the northern hemisphere was there anyone who could be called 'a duck farmer'. This is in contrast to geese, which were kept in large numbers in some localities and constituted a main livelihood for some individuals. In many parts of Europe, spring-hatched geese were driven long distances, over many days, to great autumn goose fairs, whereas ducks or duck eggs were carried to market in small numbers and in a basket.

In the Aylesbury district of England, large numbers of white-feathered ducks were raised by cottagers in the late 18th and 19th centuries. Before the advent of the railways, these Aylesbury ducks were collected and driven in large numbers down the 70 km of road to London.

Records for India and Asia do not reveal how far back specialized duck farmers can be traced, but it seems likely that the practice of producing duck eggs for sale in large numbers developed with the growth of city markets and the availability of motorized transport in the 20th century. In this system, the primary producer used his legs, or a beast of burden, to get his produce to the local market, but the secondary trader had a vehicle to take the goods to the towns and cities.

Commercial Duck Production

The origins of poultry production as a principal source of income can be traced to the development of artificial incubation in the late 19th century. The Chinese in ancient times perfected methods of incubating eggs using fermenting manure as the heat source and the French developed incubators with oil lamps in the

17th century, but it was not until well into the 19th century that factory-made incubators, with instruction manuals, could be bought for ready cash.

Incubation on an industrial scale allowed individuals with limited land to choose poultry production as their principal source of income and, in Europe and North America, this led to the development of companies specialising in supplying stock to the poultry farmer. Later, integrated businesses appeared which controlled all stages from the primary breeding programme, through multiplication and commercial production, to processing and marketing.

Although, in the early stages of industrialization, some poultry farms raised chickens, ducks, and geese and perhaps turkeys too, this soon gave way on grounds of operating efficiency and disease control to single species businesses.

Geography and Culture

The way that ducks are housed and managed varies considerably around the world, reflecting differences in climate, terrain, economic development and market demand.

Large-scale duck production in Europe and the Americas aims to sell meat in a market where consumers seek variety and can afford to pay more than the minimum price for their meat. In India and Asia, although some ducks are grown for meat, the production of duck eggs for human consumption is much more important.

In both temperate and tropical regions, the housing needed for adult ducks can be cheap, simple night shelters. Ducks need protection from predators at night and, in the colder areas, shelter from the worst of the winter weather. But, so long as they can huddle together and keep their feathers dry, ducks can withstand short periods of sub-zero temperatures perfectly well.

Where the winter is prolonged and hard, day and night housing is required for adults. These houses must be insulated and have controllable ventilation to retain heat and so keep the litter dry. The higher cost of this more elaborate housing is more than covered by the reduction in the amount of feed needed to maintain the birds.

In temperate climates, ducklings were originally produced in the spring and summer, when hatching eggs were available and simple night shelters for the ducklings were sufficient. But for year-round rearing of table duckling, well-constructed insulated houses are also needed (see Fig. 2.1). Insulation reduces the amount that has to be spent on heating during the first few weeks of life and also helps to reduce overheating of the building by solar radiation during the summer.

Use of Open Water

In many places in India and Asia, ducks are allowed access to open water during the day and are housed only at nights (see Fig. 2.2). The birds are kept indoors for 1 or 2 h after sunrise, by which time egg laying is almost completed and eggs can easily be collected after the birds are released. These flocks are

Fig. 2.1. Intensive naturally ventilated rearing accommodation.

Fig. 2.2. Tropical breeding accommodation (photographer unknown).

usually kept primarily for egg production. Surplus drakes may be killed for the pot or taken to market in small numbers, but it is the income from egg sales that is important to the family.

In some tropical areas where water is plentiful and land is scarce, the duck houses are built on stilts over the river or lake or are placed on floating pontoons, so that droppings are voided into the water. In Indonesia, Vietnam and elsewhere, this system has been developed into a fully integrated farming system with fish harvested commercially. The fertilization of the aquatic flora by the duck droppings allows obtaining a higher yield of fish. In this system, the fish may be the primary source of income, with duck eggs forming a secondary part of the business. Stocking rates up to 2000 ducks per hectare of water are reported as sustainable; beyond this, fish populations decline and the productivity of the whole system is reduced.

In many of these water-based duck farming systems, little or no feed is supplied by the farmer. Ducks maintain themselves largely by foraging, although local children may catch snakes, crabs, frogs and toads to sell to the duck farmers. The system is therefore highly efficient in terms of the ratio of costs to returns, even if the yields per breeding female are not as great as in more intensive systems.

Duck Ponds

Backyard duck keeping in Europe used to be like that too: some small return for essentially no outlay. Ducks obtained most of their food from the duck pond. Such small-scale domestic duck keeping, making use of a pond or a river, can still be found in Europe and in many other countries around the world. However, commercial duck production in Europe and North America generally makes no use of scavenging: all food used in the enterprise has to be bought and paid for. Feed conversion ratio is therefore the most important single index of success for duck production when using a fully housed system.

Some commercial duck producers, particularly in Eastern Europe, provide outdoor runs for their ducks often with access to water so that the birds can bathe (see Figs 2.3 and 2.4). This allows the ducks to choose between being indoors and outdoors, according to the prevailing weather.

In continental climates, providing access to bathing water is a valuable means of keeping ducks cool during hot weather. However, neither the area of land nor the volume of water provided in these systems is enough to support an ecosystem capable of making any significant contribution to the ducks' diet.

Rice Paddies

Many ducks are kept in association with rice paddies. Here the duck's food is supplied mainly or entirely by scavenging among the rice plants (see Fig. 2.5). The ducks help to control pests, such as water snails and insects but, to avoid damage to young rice plants, the ducks must be herded so that they graze only on paddies that are ready to take them. Herding is a good job for a small boy.

Reports from Vietnam mention that imported ducks of 'superior' genotype are no good for rice paddies because they are too lazy to graze. Thus, they have to be fed while control of pests on the paddies requires the use of chemicals.

Fig. 2.3. Semi-intensive breeding accommodation (photographer unknown).

Fig. 2.4. Table duckling on fish lake in Eastern Europe.

Hygiene

The need to prevent the spread of disease has a major effect on the system of duck keeping practised. On a small scale, ducks can be left free to roam and even to mix with other farm animals. In this case, the benefits of giving each

Fig. 2.5. Laying ducks on rice paddy. (Photograph courtesy of Tran Thanh Van, Thai Nguyen University of Agriculture and Forestry, Vietnam.)

duck a limitless volume of fresh air to breathe and a large area of territory to scavenge outweigh the risks from cross-contamination between species. But, as flock size increases and the area available to each duck is reduced, attention has to be focused on preventing the entry of disease organisms and limiting the opportunity for pathogens to spread.

Thus, large-scale duck production in developed countries makes use of tightly fenced enclosures designed to exclude all other birds. Mixing of age groups is avoided by adopting an 'all in, all out' policy for a given site. After removal of the birds at the end of a production cycle, the houses are thoroughly cleaned and disinfected. Transfer of pathogens from one site to another by humans is prevented by requiring all personnel to change outer clothing and footwear before entering the premises.

In addition to preventing the spread of disease among ducks, measures are needed to protect the human population from duck-borne pathogens, especially *Salmonella* spp. and avian influenza.

Bacteria

In a tightly controlled housing system, the main route of entry of *Salmonella* organisms is the feed. In many countries this has led to tight controls on

Salmonella in feed and the feed industry has adopted various measures to meet the stringent requirements. These include banning, or restricting by source, animal-derived ingredients such as fish meal, inclusion of materials with bactericidal properties such as sorbic acid and steam pelleting of the final product.

Strict control of hygiene at the hatchery is also required, since bacterial infections in or on a few eggs can easily spread throughout a batch of newly hatched ducklings.

Chlamydia, a fungus which can infect the oviduct causing sterility, is another pathogen that has to be rigorously controlled by good hygiene.

Avian Influenza

Waterfowl are the main natural reservoir for influenza A viruses. Duck populations in southern China are permanent hosts to a number of serotypes that live in the bird's gut, producing no symptoms in the adults. However, these viruses can mutate into more virulent forms and can spread to other hosts including pigs and man. Bird 'flu' can also multiply in other bird species and can be carried around the world by migratory birds.

Avian 'flu' viruses do not readily transfer to man, but pigs in frequent contact with bird droppings can act as hosts in which the virus adapts to the lower body temperature of the mammal (37°C as opposed to 41°C for birds), leading to subsequent infection of humans in close contact with the pigs.

Although the common avian influenza viruses do not transfer from person to person, human influenza epidemics have occurred in the past as a result of mutation of an avian virus into a form which has the capability of multiplying rapidly in humans. The possibility that a new strain could emerge, which is both virulent and highly infectious for humans, causes a good deal of anxiety amongst those responsible for public health. This in turn affects official attitudes to the siting and operation of large poultry units.

3   Housing and Environment

This chapter reviews the factors affecting the environmental requirements of growing and breeding ducks, describes methods of housing and environmental control and considers how the increasing cost of energy and concerns about bird welfare, environmental pollution and public health are likely to affect housing and environment.

History

Lee (1915) reviewed housing and husbandry for breeding and fattening ducks and described the semi-intensive system where birds were housed in simple, naturally ventilated, accommodation and allowed access to sand runs or open yards (see Fig. 2.3). He suggested that 'the aim should always be to keep ducks comfortable and avoid crowding', an advice entirely relevant today. The semi-intensive system (see Chapter 2, this volume) was to remain the principal method of production in the USA, the European Community (EC) and Eastern Europe until well after the conclusion of the Second World War. However, by about 1950, the increasing scale of production, difficulties in maintaining hygiene in open yards, along with substantial pilfering of feed by wild birds, encouraged producers in the USA and UK to consider rearing birds intensively.

Hunter and Scholes (1954), in a comprehensive review of housing and husbandry used by members of the Long Island duck producer's cooperative, located around Eastport in New York State, USA, suggested that accommodation should be insulated to prevent condensation and commented that, although no one has yet worked out ventilation requirements for the duck, pure fresh air is the 'cheapest feed you can give to growing ducks'. They also suggested that electric fans should be used 'to provide fresh air and remove foul moisture-laden air'.

From about 1965, duck producers in the USA and UK began increasingly to rear birds intensively at about 4–5 birds per m^2 in wide span insulated buildings, supplied with fresh air by pressure jets and ventilated with electric fans. However, although housing, methods of heating and ventilation have been 'borrowed' from the broiler industry, data presented in this chapter will demonstrate conclusively that the domestic duck is not just a waterproof chicken.

Eastern Europe and Russia

These countries have a long tradition of duck production integrated with agriculture and aquaculture. The Pekin is widely used principally because of its rapid growth rate and white feathers and in particular its white down, which is a highly valued by-product and is of exceptional quality where birds are grown with access to water. Many birds are reared semi-intensively with access to range or fish lagoons (see Fig. 2.4). Because of the severe winters, production of table duckling is limited to the period from April to October when open water is available. Breeding ducks are normally housed in daub and wattle or wooden buildings with thatched or asbestos roofs and natural ventilation. Outside yards are provided, with a water channel or limited access to fish pond, lagoon or canal.

Breeding stocks are usually hatched in June and July and brooded using either gas or hot-water brooding systems for about 2 weeks and given access to an outside run and bathing channel. Wood shavings, sawdust or rice hulls are used as bedding material during brooding, compounded feed is provided from day-old and feed restriction is practised from about 3–4 weeks of age when birds are transferred to either grass range or an area of previously cropped land and given access to canals or fish lakes. At about 16 weeks birds are moved from the range to their breeding quarters. Natural lighting is used from day-old until artificial light is provided in late December to bring birds into lay in late January at about 28 weeks of age. The first eggs for commercial hatching are set in mid February, hatched in mid March, brooded (see Fig. 3.1) and transferred to range or fish ponds (see Figs 3.2 and 2.4) at about 24 days where they continue growing until slaughter at between 7 and 8 weeks of age. Eggs are set weekly until late August and the last commercial table ducklings grown on range are slaughtered in mid October before the onset of winter.

Confining breeding stock to their accommodation until about 08:30 h encourages birds to use nests and protects hatching eggs from low winter temperatures. Food is provided *ad libitum* and artificial light, both inside the breeding accommodation and outside, provides more than 16 h daylength. After the last egg collection (about 08:30 h) birds are excluded from the accommodation and given access to outside yards.

European Community

In the EC, and increasingly in parts of Eastern Europe, production of table duckling is a year-round business. Breeding stock and table duckling are

Fig. 3.1. Semi-intensive brooding accommodation with access to water.

Fig. 3.2. Rearing table duckling on open range with access to water for bathing.

normally housed in intensive, insulated power-ventilated accommodation, which is similar in most respects to the buildings provided for other types of poultry, but with drinkers located over slatted, plastic or wire flooring drained to an effluent disposal system. However, nipple drinkers (see Fig. 3.3), plastic turkey drinkers or modified pig troughs (see Fig. 3.4) located over some form of floor drainage to an effluent disposal system can provide drinking water to both table ducklings and breeding ducks to eliminate the cost and the increasing welfare concerns associated with slatted or plastic flooring.

Fig. 3.3. Nipple drinkers.

USA and Canada

Long Island in the state of New York has long been synonymous with rearing ducks semi-intensively. Birds are given increasing space with age and allowed limited access to sand runs. Commercial production of table duckling commenced about the beginning of the 20th century and by 1950 (Scott and Dean, 1991) there were more than 40 commercial duck farms located on Long Island. In 1953 the Long Island Duck Growers Marketing Cooperative provided finance for construction of a Duck Research Laboratory at Eastport, Long Island. Research was jointly undertaken by Cornell University and the Long Island Research Cooperative and contributed significantly to growth and development of the duck industry worldwide.

Producers on Long Island reared ducks on earth floors covered with straw bedding in insulated and naturally ventilated accommodation between 5 and 12 m wide, with sidewalls between 1.5 and 3 m high. Some producers provided slatted flooring over concrete pits connected by drainage to an effluent disposal system. Birds were given access to sand-covered or concrete yards located along one side of the accommodation. Water was provided in a ditch or cement trough (see Fig. 2.3). However, increasing federal legislation affecting disposal of effluent along with rising real estate values contributed to a gradual decline in duck production.

Production systems used in Canada and the northern states of the USA are broadly similar to those previously described for the EC, with the exception that some producers rear table duckling on slatted plastic flooring (see Fig. 3.5). One significant difference is that fresh air is sometimes supplied to both rearing

Fig. 3.4. Waste-free open water trough.

and breeding accommodations through pressurized perforated tubular plastic ducts extending along the length of the building (see Fig. 3.6). During late autumn and winter, incoming air is heated and mixed with air within the building and distributed via plastic ducts to maintain environmental temperature.

In the southern USA and Brazil, table ducklings are reared in naturally ventilated, insulated accommodation about 20 m wide with sidewall curtains (see Fig. 3.7) which can be raised or lowered according to weather conditions. Brooding is confined to one end of the building by a curtain or movable partition. Birds are reared on wood shavings, and nipple drinkers are used to reduce wastage and subsequent effluent. Overhead fans are widely used during the summer months to provide air movement over the birds.

China

The semi-intensive system of housing using naturally ventilated simple accommodation of local materials (bamboo, timber and concrete blocks) is widely

Fig. 3.5. Intensive rearing on plastic-coated weld mesh flooring.

Fig. 3.6. Intensive breeding accommodation with fresh air supplied through pressurized tubular ducting.

used for housing breeding stock. Birds are usually given access to outside pens and bathing water (see Fig. 2.2), because many producers consider it essential to provide swimming water to encourage fertility. In tropical and subtropical regions, overhead shade in the form of thatched cover is provided and birds are given confined access to swimming water or canals.

Fig. 3.7. Intensive naturally ventilated rearing accommodation with adjustable weatherproof sidewall curtains.

An extensive system is commonly used for the production of table duckling. The ducks are continually tended while they graze over rice paddy (see Fig. 2.5), where birds perform an important function of scavenging for insects while they forage for whatever food is available. The birds are given a feed of cereal grain in the evening and are housed overnight to protect them from predators. Eggs are laid in the night before the ducks are released in the morning. In the eastern and southern provinces of China, ducks are reared on the seashore and river deltas, where a similar system of pasturing along the seashore provides the majority of their food.

In recent years the importation of Pekin from the EC and the USA has led to the introduction of intensive systems for production of both hatching eggs and table duckling, using systems similar to those previously described. However, breeding birds are still sometimes reared by pasturing over rice crops until approaching sexual maturity, when they are transferred to intensive breeding units.

The Far East

Traditional production of both hatching eggs and table duckling in most countries of the Far East is similar to that in China. However, in Burma, Vietnam and Indonesia the whole production cycle is sometimes carried out on water. Edwards (1985) describes how integration of fish and duck farming can substantially improve productivity and reduce economic risk, since two sub-systems are involved as opposed to a single-commodity farming system.

Duck sheds are usually simple shelters constructed of bamboo, thatched with palm leaves providing shade and protection against heavy rain. Shelters are frequently built over ponds or canals or may be floated on water. This allows fish to consume waste feed and effluent.

In recent years there has been a very rapid growth in intensive duck production in many countries in the Far East. Increasingly birds are reared in naturally ventilated, open-sided accommodation provided with insulated roofing and sidewall curtains which can be adjusted according to weather conditions. Birds are frequently given access to bathing water. Overhead fans and evaporative cooling are sometimes provided to ameliorate the effects of high environmental temperature.

Farm Size

There does not seem to be any definable upper limit to the number of ducks that can be housed on one site, but all the empirical evidence indicates that performance is never as good on multi-age farms as on sites operated with an 'all in, all out' policy. Commercial trials on multi-age sites, investigating increased hygiene and disinfection or improved environment or husbandry practices, have shown little improvement in performance or mortality rates. This implies that the adverse effect of rearing mixed age groups near to each other is due to unidentified, but transmissible, low-grade pathogens.

For large-scale duck production, the additional investment in land and transport costs involved in providing separate sites for each age group is more than justified by the better performance obtained. An additional advantage is that regular comparison of performance at the different sites provides a powerful tool for identifying and correcting any problems (see Process Control, Chapter 4, this volume).

House Size

The dimensions of both rearing and laying accommodation are governed by climate, cost and tradition. For example, in hot and Mediterranean climates the width of open-sided, naturally ventilated accommodation is normally restricted to about 10 m to assist cross ventilation, but in the EC and the USA, dimensions of intensive, power-ventilated accommodation are principally governed by cost per square metre of effective floor space along with local building legislation and planning controls.

Sidewalls in both slatted and littered accommodation in hot and cold climates must provide sufficient headroom to allow working staff to carry out routine husbandry tasks and permit mechanical removal of litter. Providing ample headroom along sidewalls also assists with controlling temperature in cold climates, because increasing internal volume dilutes the effect of cold airflow on environmental temperature.

Slatted Flooring

Slatted flooring in part litter and slatted accommodation should be located to allow for a gradual build-up of litter. Slatted platforms should be about 30 cm above floor level (assuming a ratio of 70:30 for litter and slatted floor areas) for table duckling placed at 6 birds per m². For parent stock housed at between 2 and 3 birds per m² the slats should be about 130 cm above the floor. Profiled timber and weldmesh are both widely used as slatted flooring, but moulded plastic also provides a cost-effective, hygienic and comfortable surface for birds to walk and rest on. Slatted flooring must be located over channels with sufficient fall to encourage drainage to either an approved drainage system or an effluent storage facility. Drainage systems can become obstructed with effluent and so it is prudent to avoid locating drainage channels under concrete flooring and roads. It is also essential to provide service access to all channels, gullies and underground drains. Effluent pits and canals should be constructed and protected as required by both building and health and safety regulations.

Provision for Cleaning

It is essential to design hygiene into any proposed accommodation. Concrete floors should provide a smooth surface with sufficient fall to facilitate efficient drainage to gullies covered with perforated metal or plastic, located along either sidewalls or the midline of the building. Dwarf walls of poured concrete with a concave surface joining the wall to the floor will prevent a build-up of debris and provide a smooth surface for cleaning and disinfection.

Fan shafts, air inlets and internal surfaces should be lined with durable, smooth and waterproof material to facilitate cleaning and disinfection. Roof trusses, brackets and stanchions should be similarly protected to prevent build-up of dust and debris. Locating bell mouth extractor fans in removable panels, connected to the electric supply with detachable waterproof plugs, facilitates removal for effective cleaning and disinfection of fan shafts and also replacement in the event of failure.

After depletion, all litter should be removed and the electric supply disconnected. Power for lighting and washing equipment should be obtained from a separate source. It is advisable to 'waterproof' all electric fittings to prevent water penetrating and damaging electric and electronic circuits. All internal surfaces should be washed with clean water followed with an approved disinfectant solution at the correct concentration. Government health and safety at work regulations must be followed when operating electrically powered washing equipment and gas-fired heating equipment.

Other Considerations

To satisfy welfare regulations and prevent panic in the event of power failure it is essential to use an interval timer to provide birds with a period of darkness each day. Light intensity can be controlled using either an electronic dimmer or low-energy fluorescent bulbs located at 12 m intervals. It is also advisable to supply a separate lighting circuit, controlled by a proximity switch, to provide sufficient light for staff to carry out their routine tasks and inspections.

It is essential to protect birds from predators, including thieves. Weasels, stoats and snakes are adept at gaining entry and leave little evidence of their activity. Staff should carry out regular inspections both inside and outside the accommodation and seek advice and assistance from a reliable pest control service in the event of predation or infestation.

Feed in many countries is delivered in bulk and stored in feed bins. It is advisable to store starter and grower feed in separate bins. It is essential to provide water storage in each house with sufficient capacity to supply drinking water for at least 8 h to cover any temporary failure in main water supply.

Temperature

Heat output from metabolic activity is affected by genotype, age, feed intake and physical activity. To maintain heat balance and body temperature, heat loss must equal heat production. In cool temperate conditions, birds use mainly non-evaporative heat (sensible heat) transfer by convection, conduction and thermal radiation to maintain heat balance, but the ability of birds to lose heat in this manner declines with increasing temperature, diminishing to zero as environmental temperature approaches the body temperature ($41°C$).

When temperature increases beyond the upper boundary of the thermo-neutral zone (see Fig. 3.8), birds achieve heat loss and maintain homeostasis by increasing evaporative heat loss (insensible heat loss) through latent heat exchange during respiration. Inspired air passes over the wet mucosal surface of the upper respiratory tract, cooling blood in the underlying tissue, and air is expired at a higher temperature and humidity. Above about $26°C$ ducks substantially increase rate of respiration to optimize evaporative heat loss, but use shallow breathing (gular flutter) to avoid overventilating the alveoli in the lungs (see Fig. 3.9), thus preventing excessive depletion of blood bicarbonate. Vaporizing moisture during respiration to achieve heat loss depends on the difference in water vapour pressure (mass of water vapour per unit volume) between the bird's air passages and the environment. At high environmental temperature, the difference in water vapour pressure is more important than dry bulb temperature.

Figure 3.8 describes the relationship between environmental temperature and rate of heat production for mature duck fed *ad libitum*. Empirical experience in both Eastern Europe and the USA shows that mature, fully feathered birds can maintain thermoregulation over an environmental deep shade dry bulb temperature range between about $-8°C$ and $40°C$, provided the air is relatively dry.

Mature acclimatized ducks exposed to dry bulb temperatures below about $-8°C$ for sustained periods are unable to produce sufficient heat to maintain their body temperature and gradually become hypothermic and die. Conversely, acclimatized birds maintained in deep shade exposed for several hours to dry

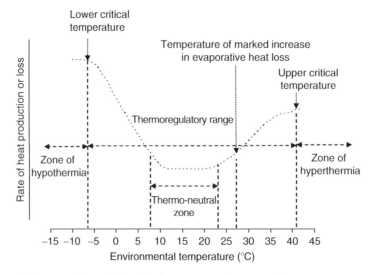

Fig. 3.8. Diagram of the relationship between environmental temperature and heat production and loss for *ad libitum* fed mature Pekin with a bodyweight of about 3.6 kg based on empirical evidence from Eastern Europe and North Carolina, USA. (Diagram modified from Mount, 1979.)

Fig. 3.9. Table duckling shallow breathing to assist heat loss.

bulb temperatures above about 40°C are unable to lose metabolic heat and experience an increase in deep body temperature and become comatose and suffer heart and respiratory failure. However, depending on environmental conditions, ducks can modify their behaviour and extend their thermoregulatory range. In northern latitudes birds housed semi-intensively tolerate low daytime temperatures in outside pens by occasionally clustering together into 'rafts' to reduce heat loss and collectively 'warming up'. In the hot and wet tropics access to bathing water allows birds to increase sensible heat loss and so tolerate high radiant temperatures and increased water vapour pressure.

The 'thermo-neutral zone' (see Fig. 3.8) describes a temperature range of minimal metabolism. When environmental temperature falls below or rises above the zone boundaries, ducks either use energy to produce heat or increase respiration to lose heat, achieve heat balance and control their body temperature. However, although environmental dry bulb temperature is widely used, both to describe and to monitor the environment, rearing birds on composted straw and similar materials (where litter temperature usually exceeds about 28°C) reduces their sensible heat loss at low environmental temperature, but conversely increases heat load and the respiration rate required to maintain heat balance at high temperatures. Rearing birds on fermented litter reduces both their lower and upper critical dry bulb temperatures and shifts the thermo-neutral zone compared to birds reared on slatted flooring. Conversely, in hot climates, air movement through slatted or bamboo flooring assists birds to lose heat, but increases energy requirement to maintain homeostasis at low temperature. Other factors, such as radiant temperature, air speed, water vapour pressure and controlling feed intake, can also affect heat balance.

Increasing environmental temperature above the lower boundary of the thermo-neutral zone reduces feed intake (see Fig. 4.4) and subsequent rate of gain (see Fig. 4.6), while reducing temperature below the lower boundary adversely affects feed conversion and inhibits litter fermentation, affecting both feathering and health. This means that in hot and Mediterranean-type climates optimum feed intake and rate of gain for a given genotype and feed can be achieved by:

- Supplying accommodation designed to prevent solar radiation and heat from fermenting litter from increasing sensible heat load on the growing bird;
- Providing sufficient ventilation to maintain environmental temperature close to ambient deep shade dry bulb temperatures along with ample air movement to allow birds to optimize sensible and insensible heat loss.

Conversely, achieving optimum growth and economy of feed conversion in cold and cool temperate climates depends on maintaining environmental temperature just above the lower boundary of the thermo-neutral zone by:

- Providing well-insulated and windproof rearing accommodation along with either time- or speed-controlled electric fans to supply minimum ventilation, conserving heat from birds and litter;
- Either rearing birds on fermenting litter to provide a comfortable bed and warm environment or providing artificial heat if birds are reared on slatted flooring.

Ventilation

Effective ventilation requires:

- Sufficient ventilation to reduce pollutants and gases below an acceptable limit by gradually increasing minimum ventilation in temperate climates from about 20% to 40% of maximum ventilation between 17 days and age at slaughter (see Fig. 4.34);
- Precise control of minimum ventilation along with sufficient insulation ($U = 0.4\,W/m^2.°C$) to prevent unnecessary heat loss in cold weather;
- Sufficient maximum ventilation to maintain house temperature <2°C above ambient temperature immediately prior to slaughter in hot weather by providing $1.2\,m^3/s$ for each 1000 kg body mass for birds reared on a ratio of 70:30 litter and slatted floor;
- Windproof air inlets and outlets (see Figs 3.10 and 3.11);

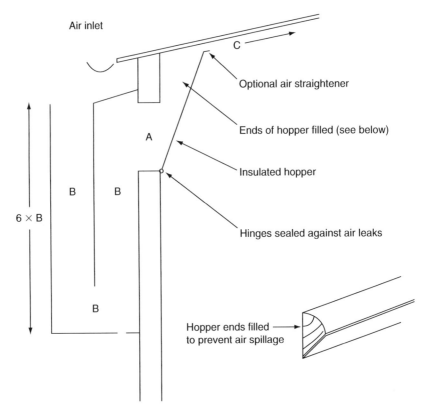

Fig. 3.10. A typical light and wind baffled inlet. Dimensions A, B and C require detailed calculation, but typically result in air speeds of 3.5, 4 and 5 m/s, respectively. (Reproduced from Charles *et al.*, 1994a; Fig. 9.6. With permission from CAB International.)

- Air recirculation when brooding to prevent stratification of temperature and provide a uniform environmental temperature to prevent crowding and increase effective space per bird;
- Fail-safe equipment to provide sufficient natural ventilation in the event of a failure of electric supply to maintain bird comfort and welfare.

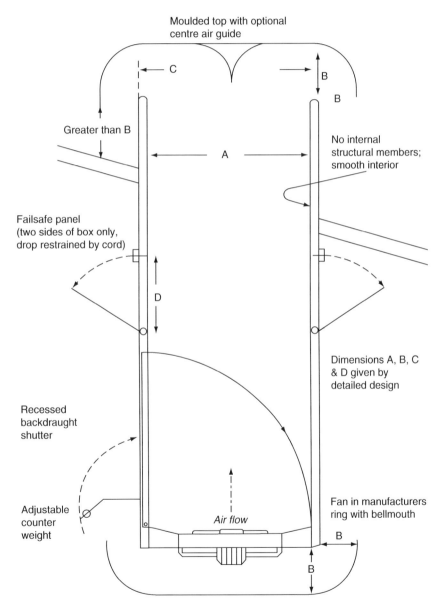

Moulded top with optional centre air guide

C

B

B

Greater than B

A

No internal structural members; smooth interior

Failsafe panel (two sides of box only, drop restrained by cord)

D

Dimensions A, B, C & D given by detailed design

Recessed backdraught shutter

Adjustable counter weight

Air flow

Fan in manufacturers ring with bellmouth

B

B

Fig. 3.11. Fan box with fail-safe. (Reproduced from Charles *et al.*, 1994b; Fig. 9.5. With permission from CAB International.)

Wind can profoundly affect ventilation in both naturally and artificially ventilated buildings. Pressure on sidewalls and end walls and along either side of the
roof varies and is affected by location, dimensions, roof pitch and both wind
speed and direction. Pressure increases with wind speed, profoundly affecting
the efficiency of air inlets and outlets. Empirical experience using wide buildings
with low-pitch roofs for rearing table duckling shows that wind can exert sufficient negative pressure to prevent ridge-mounted extractor fans, even operating at full speed, from functioning efficiently. Wind speeds above about force
6 (Beaufort scale) can reverse airflow with air entering through exhaust fans
and exiting leeward, via sidewall air inlets. Ventilation of windproof buildings
can be achieved using electric fans to either increase or reduce air pressure
within the building. However, the distribution and pattern of air movement
within a building, and air speed over birds are affected by the location and
design of air inlets.

Ventilation in Cold and Cool Temperate Climates

In cold and cool temperate climates, limiting ventilation in insulated accommodation allows the use of metabolic heat to maintain house temperature. Figure
3.12 shows that temperature can be increased by about 1.2°C for each reduction of 10% below 80% of total ventilation. However, productivity is also
affected by bird health, and birds housed intensively in closed accommodation

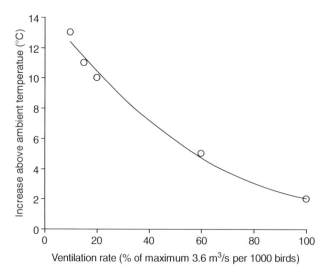

Fig. 3.12. Effect of reducing ventilation on increase in environmental temperature
above ambient dry bulb temperature. Measurements were recorded in insulated
(0.52 W/m². °C) intensive rearing accommodation with 70% litter and 30% slatted
floor holding about 6500 birds aged about 30 days placed at 6 birds per m² when
ambient temperature was between 3°C and 6°C.

must have sufficient ventilation to control both temperature and moisture, and to remove microorganisms, dust, waste gases, fumes and odours.

Empirical experience, along with performance measured in large-scale commercial trials, confirms that the following methods of ventilation are relatively windproof and suitable for rearing table duckling placed at about 6 birds per m^2 given litter (70%) and slatted flooring (30%) from about 14 days to slaughter in insulated ($U = 0.5 \text{W/m}^2 \cdot °C$) accommodation in cool and temperate conditions.

1. Sidewall air inlet and side wall extraction

Fans located in the roof or along sidewalls provide sufficient negative pressure (50 N/m²) to encourage fresh air to enter the accommodation through narrow sidewall inlets (see Fig. 3.13a) located along the eaves and extending the length of the building. Air travelling at high speed (about 5 m/s) follows the internal surface of the roof lining (Coanda effect), but gradually slows down and mixes with internal air. The cool mixture then gradually sinks, providing birds with a draught-free supply of fresh warm air.

2. Pressure jet air inlet and sidewall extraction

Supplying sufficient ventilation to maintain relative humidity below 85% encourages evaporation of moisture and composting activity in litter, and provides birds with a warm dry bed, profoundly affecting their comfort, welfare, feathering and health. However, at low temperature (<0°C), air flowing through sidewall inlets cools internal air below the 'dew point' (the temperature at which water vapour starts to condense when air is cooled), occasionally even producing dense fog. The cool, saturated air rapidly sinks to floor level and causes crowding, which reduces effective floor space and litter becomes saturated with moisture (>16% water) and faeces, affecting health, feathering and physical performance. However, commercial trials in the EC and Eastern Europe at low ambient winter temperatures (<0°C) confirm that pressure jets provide an effective and economic method for avoiding these problems by mixing cold fresh air with warm internal air and distributing the mixture uniformly throughout the accommodation.

Fans located in either one or both end walls (see Fig. 3.13b) supply fresh air distributed under pressure through perforated plastic tubing with a diameter of about 600 mm (total area of holes between 1.5 and two times the cross-sectional area of plastic duct) located about 2 m above floor level and extending along the length of the building (see Fig. 3.6). Wind-protected electric fans located in the roof or along sidewalls, controlled by interval timer and thermostats, reduce the pressure in the building. Fresh air is drawn into the plastic ducts and distributed evenly through the building. At low temperature, reduced ventilation provides sufficient fresh air to maintain air quality and constant recirculation provides birds with an even temperature and draught-free environment.

3. Positive pressure through porous ceiling

Fans located in the roof increase air pressure in an enclosed roof attic space and supply fresh air at low speed (between 0.1 and 0.2 m/s) through a porous

fibreglass (50 mm) or perforated plastic membrane ceiling (see Fig. 3.13c). The
fresh air cools the internal air and the mixture gradually descends to floor level
with positive pressure, encouraging air to egress through sidewall outlets fitted
with backdraught protection.

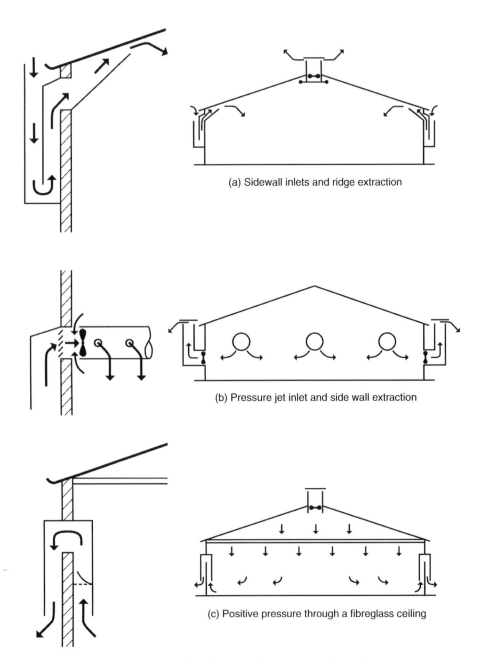

(a) Sidewall inlets and ridge extraction

(b) Pressure jet inlet and side wall extraction

(c) Positive pressure through a fibreglass ceiling

Fig. 3.13. Pattern of air intake and flow for alternative systems of ventilation.

4. Natural ventilation

Heat from birds and fermenting litter increases air temperature, reduces air pressure and increases air buoyancy; warm air rises (stack effect) and encourages entry of fresh air through inlets located along sidewalls and the escape of warm stale air through outlets located along the ridge of the roof.

Air pressure and subsequent ventilation are affected by the difference between ambient and internal temperatures, along with the vertical distance between air inlets and outlets. Locating air inlets either at floor level or below floor level for birds reared on litter and slatted flooring, respectively, assists 'stack ventilation', and either manual or mechanically adjustable shutters can be used to control airflow and ventilation. Alternatively, Yorkshire boarding (profiled timber boards with gaps between them) located along sidewalls in widespan buildings (see Fig. 2.1) can also provide excellent natural ventilation.

Ventilation in Mediterranean and Hot Climates

In Mediterranean and hot climates ducks of all ages are usually housed in open-sided accommodation provided with steep pitch roofing and an open ridge to encourage stack ventilation (see Fig. 3.14). Adjustable sidewall curtains (see Fig. 3.7) are provided where there is seasonal variation in either temperature or rainfall.

Local geography and orientation to prevailing wind substantially affect natural ventilation. Rearing birds on slatted flooring in accommodation less than 8 m wide, and with the long axis across the direction of the prevailing wind, either on hillsides or on stilts over water encourages both stack effect and cross ventilation.

Where birds are reared in closed houses evaporative cooling can reduce air temperature and heat stress (see Chapter 4, this volume), but this can only be achieved at the expense of increasing water vapour pressure, reducing the opportunity for birds in hot humid climates to lose insensible heat through

Fig. 3.14. Naturally ventilated intensive duck-rearing accommodation.

respiration. However, in a Mediterranean climate with relatively low water vapour pressure, distributing a microsol of water with fans (see Fig. 3.15) or through a pressure jet system to prevent litter becoming damp (see Fig. 3.16) can reduce daytime dry bulb temperature by as much as 8°C.

Fig. 3.15. Naturally ventilated intensive rearing accommodation provided with large-diameter fans used to distribute a microsol of water and reduce environmental temperature.

Fig. 3.16. Naturally ventilated intensive breeding accommodation provided with pressure jets to distribute a microsol of water to reduce environmental temperature.

Producers in the southern USA use 'tunnel ventilation' to alleviate heat stress in broilers during summer, and observation of table duckling and breeding stock in both the Far East and during summer in North Carolina confirms that ducks enjoy resting in a stream of fast-moving air. However, unreported commercial-scale trials confirm that increasing air speed to 3 m/s over birds, using either tunnel or cross ventilation, or providing large volumes of air at relatively slow air speed with ceiling fans, had no measurable effect on performance, but encouraged crowding, because birds congregated in the stream of moving air, adversely affecting litter and subsequent feather quality.

Ventilation in a Continental Climate

Achieving optimum economic performance in a continental climate depends on maintaining environmental temperature above the lower boundary of the thermo-neutral zone during winter and preventing internal temperature exceeding deep shade dry bulb temperature by more than 2–3°C during summer.

Rearing birds on fermenting litter provides sensible heat and assists towards maintaining environmental temperature during winter but, conversely, substantially increases heat load and environmental temperature during summer. Table ducklings in Eastern Europe are reared on litter in winter, but are given access to outside pens and bathing water to reduce heat stress during summer. However, legislation affecting disposal of effluent along with increasing concerns about transmission of disease from migrating wildfowl to domestic duck (and the human population) limit opportunities to rear birds in this manner. Increasingly, table duckling in continental climates are reared in insulated and power-ventilated accommodation on either slatted plastic flooring or a combination of litter and slatted flooring to avoid or reduce heat from fermenting litter affecting environmental temperature during summer.

In cold weather, birds reared on slatted flooring require artificial heat to maintain internal temperature above the lower boundary of the thermo-neutral zone, but unreported commercial trials confirm that providing litter over about 50% of the total floor area provides birds with a warm bed and sufficient sensible heat to maintain internal temperature about 9°C above outside dry bulb temperature. However, providing both litter and slatted flooring will affect behaviour, because birds congregate and rest on warm litter in winter, but prefer slatted flooring during the heat of summer. It is important to provide sufficient space to avoid crowding and physical damage by adjusting number of birds placed according to genotype, area of slatted flooring and time of year.

During winter, birds reared on slatted flooring or a mixture of litter and slatted flooring require artificial heat until about 26 and 20 days, respectively.

Limiting ventilation avoids wasting energy, but it is essential to provide sufficient fresh air to maintain air quality (NH_3 <15 ppm and humidity <80%) and health. Empirical experience in cold weather confirms that recirculating air using either ceiling fans or pressure jets prevents temperature stratification and substantially reduces heating costs; distributing heat evenly over the entire floor area prevents crowding to 'hot spots' and improves feathering and welfare.

A maximum ventilation rate of $1.2\,m^3/s$ for each $1000\,kg$ body mass should be sufficient to prevent internal temperature exceeding ambient deep shade dry bulb temperature by more than about 2–3°C. Windproof air inlets (see Fig. 3.10) should be located to provide an even distribution of fresh air over the floor area. Providing supplementary air inlets below slatted flooring can increase airflow and assist heat loss from birds resting on the slatted area in hot weather.

Light baffles, back draught shutters and similar modifications can increase static pressure and substantially reduce the efficiency of electric fans because airflow is governed by static pressure. It is important to provide sufficient inlet area to prevent static pressure exceeding the fan manufacturer's recommendations (usually $50\,N/m^2$ or about 5 mm water gauge). Air inlet area can be adjusted to control air speed; about 0.15 and $0.3\,m^2/m^3/s$ of ventilation capacity are normally sufficient to achieve inlet air speeds of 5 and $2.5\,m/s$, respectively.

Insulation in Cold and Cool Temperate Climates

Heat transfer through a building structure is governed by the difference between internal and external surface temperatures and the thermal properties of the structure. Insulation reduces the conduction of heat flowing through a structure, reducing heat loss in cold weather and heat gain in hot weather.

In cold or cool conditions about 70% of metabolic heat loss from birds is used to warm minimum airflow and maintain environmental temperature above the lower critical temperature; the remaining 30% of metabolic heat is lost through the structure. Increasing dry bulb temperature substantially increases the water-holding capacity of air, allowing moisture in respired air and faeces along with water lost during bathing to evaporate. However, all internal surfaces must be insulated to prevent condensation, because moisture in warm air will condense on surfaces below the dew point.

Commercial trials in a cool temperate climate found that, for ducklings maintained at about 6 birds per m^2 on litter and wire, time-controlled minimum ventilation to keep ammonia below 10 ppm and relative humidity around 80% and roof and wall insulation values of either 0.52 or $0.39\,W/m^2$.°C could maintain environmental temperature about 9°C and 12°C, respectively, above ambient dry bulb temperature in winter. Well-feathered ducklings can tolerate night-time temperatures below freezing, but temperatures below about 8°C adversely affect litter fermentation. The litter becomes wet and dirty, which affects feathering, health and welfare.

Optimum insulation for brooding accommodation is determined by the cost of insulation relative to the cost of energy required to maintain environmental temperature in a particular climate. However, measuring the cost/benefit ratio is complex because increasing insulation reduces heat loss in a curvilinear manner and the anticipated relationship is also affected by the expected difference between internal and ambient temperatures. Charles *et al.* (1994c) suggest an insulation value (thermal conductance) of $U = 0.4\,W/m^2.\,°C$ for intensive and heated broiler housing in the UK, but it may be prudent to regard this as a minimum value because cost of fuel is likely to increase over the estimated lifetime of new accommodation.

Insulation in Hot Climates

Dry bulb temperature is widely used to measure environmental temperature, but radiant and wet bulb temperatures may provide a better description of 'effective temperature' in hot conditions. Unreported trials from the USA and the Far East confirm that deep shade radiant temperature recorded daily at about midday in summer in curtain-sided rearing accommodation with insulated roofing and extended eaves was consistently more than 2°C above the dry bulb temperature.

In hot, dry and Mediterranean-type climates galvanised iron, aluminium, asbestos and clay tiles are all widely used as roofing materials for both breeder and table duckling accommodation, but in full sun these materials can reach temperatures >50°C, exposing birds to high radiant heat loads. Entry of radiant heat to the building can be substantially reduced either by providing about 20 mm of insulation or by spraying a fine mist of water over the roof surface at regular intervals (about 10–15 min in full sun). Painting external surfaces of the walls and roof white will reflect solar radiation, thus reducing surface temperature and transfer of radiant heat.

In the wet tropics, palm leaf thatch is widely used as roofing material, providing low-cost protection from heavy rainfall along with effective insulation from solar radiation.

The Brooding Environment

Empirical experience in the EC, Eastern Europe and the Far East confirms that table duckling should be reared in clean and single-age accommodation located well away from all other poultry and wild birds, because multi-age farms provide a reservoir of pathogenic microorganisms, affecting health, physical performance and processed quality.

Brooding accommodation, drainage channels, air inlets, fan shafts and all furniture should be dusted, thoroughly washed and disinfected with an approved disinfectant at the recommended dilution. Brooders, furniture and equipment should be arranged to optimize bird comfort and encourage feed intake rather than achieving an orderly appearance and facilitating routine husbandry.

Many integrated production companies routinely carry out a bacterial examination following cleaning and disinfection to monitor hygiene. Small producers might consider arranging for a local veterinarian to regularly visit the accommodation to review overall hygiene and give advice on vaccination and similar matters. Only essential personnel should be allowed access to brooding and rearing accommodation. Everyone should wear protective clothing and footwear and wash their hands prior to entry.

Litter materials such as green wood shavings and sawdust, rice hulls and chopped straw often contain more than 12% moisture on delivery. However, spreading litter and supplying both heat and ventilation for about 24 h will reduce litter moisture content to about 3%, an essential prerequisite for providing day-old ducks with a warm, dry and comfortable bed. It is good practice to store litter under cover to prevent it becoming wet or contaminated with wild bird faeces. Fresh and dry litter should be spread daily from about 3 days to maintain a clean bed. Litter can be stored on suitably strengthened tables within the accommodation to make the task of spreading litter easy.

Brooding

Restricting the initial brooding floor area with an internal partition such as a plastic curtain increases the metabolic heat produced per square metre of floor area and assists in maintaining environmental temperature in winter.

Both zone and whole-house heating are widely used for brooding duckling. Zone heating relies on canopy brooders to supply radiant heat. Temperature at floor level directly under the canopy can exceed 45°C, but declines in a curvilinear manner with distance (see Fig. 4.32), allowing chicks to select their comfort zone. Supporting brooders with safety chains from an overhead track allows the operator to alter the distance between brooders and adjusting the height affects spread of radiant heat.

Thermostats control temperature by supplying an electronic signal to either a solenoid or a modulating gas valve, increasing or decreasing the flow of gas to brooders. Solenoid control provides either high or low radiant heat. Gas valves, by contrast, modulate gas flow, providing a continuously variable supply of radiant heat. Radiant heat output is affected by design and gas pressure. Optimum heating efficiency can be achieved by adjusting gas pressure (with all brooders operating at high flame) to provide the manufacturer's recommended pressure.

Large-scale commercial trials confirm that providing a common thermostat to control all the brooders in a house, rather than individual brooder thermostats, reduces temperature variation and also cuts labour and heating costs. It is important to adjust the differential between switching on and switching off the common brooder thermostat to about 2°C to prevent brooder temperature undulating.

Whole-house heating of brooding accommodation is widely used by commercial producers in both North America and the EC. Birds are often brooded and then moved or given access to adjacent carry-on rearing accommodation between 12 and 17 days of age. Gas-fired heating controlled by a thermostat and distributed through pressure jets prevents stratification and provides an even temperature at floor level, but farm staff are responsible for both regulating and reducing temperature with age. However, some producers provide radiant brooders, allowing chicks to select their own comfort zone, but still supply heated air through pressure jets to maintain a stable background temperature away from the brooders, thus increasing effective floor space and the number of chicks that can be placed in the house.

Air speed at floor level profoundly affects comfort and behaviour. Chicks (<6 days) crowd together when air speed exceeds about 0.2 m/s, occasionally scrambling over one another in an attempt to reduce their exposure to air movement. Chicks then become streaked with moisture and faeces, affecting growth and development of down clusters, particularly over their flanks and lower back. This substantially reduces both the quantity and quality of down recovered after slaughter.

Trials have confirmed that environmental dry bulb, radiant and litter temperature along with air speed and humidity all affect comfort, behaviour and performance, and this confounds attempts to describe environment in terms of a single parameter or as a 'requirement'. Evidence gained using diverse genotypes in different climates confirms that behaviour (appearance, activity and sound) provides a more reliable guide to chick comfort and welfare than any physical measurement. However, farm staff must be trained to recognize when environment is adversely affecting chick behaviour and physical appearance. For example, low night-time outside temperature and strong wind can affect both house temperatures and air speed. The following day chicks may appear comfortable and behave quite normally, but on close inspection show the characteristic wet streaks on down over their flanks and back caused by crowding the previous night.

Brooding in Hot Climates

Chicks in hot and Mediterranean climates are generally reared in curtain-sided accommodation and Fig. 3.17 shows a typical brooding layout. At high ambient

Fig. 3.17. Intensive naturally ventilated brooding accommodation.

temperatures, only minimal heat is required. In the Far East and during summer in Eastern Europe chicks are frequently given access to outside pens and provided with shallow bathing water (see Fig. 3.18) from as early as 3 days of age, but depth of bathing water should be limited to about 5 cm to prevent chicks from drowning. However, ducklings soon apply oil from their preen gland and waterproof their down so that by the time they are 10 days old the depth of water can be increased to >10 cm to allow chicks to swim and socialize on water.

Ventilation in the Brooding Stage

The dimensions, design and location of air inlets will affect negative pressure, the pattern of airflow and air speed in fan-ventilated accommodation. Smoke candles designed for the purpose can help to show the effects of inlet location and air speed on subsequent airflow into and within the house.

Reducing inlet area of fan-ventilated accommodation increases both negative pressure and air speed, but reduces airflow. It is essential to supply sufficient inlet area to prevent negative pressure at maximum ventilation (all fans at maximum speed) from exceeding manufacturers' recommendations (usually 50N/m^2, or about 5 mm water gauge) because reducing the air volume passing through fans increases motor temperature, affecting both lifespan and safety.

Airflow can be controlled either by reducing fan speed or by limiting the time that fans operate at full speed. Fan speed can be reduced by decreasing voltage, and modern control units provide the facility to control both minimum and maximum fan speed and airflow. However, restricting fan speed reduces air speed entering the accommodation and allows cool fresh air to sink rapidly to floor level, affecting both chick comfort and behaviour. Wind

Fig. 3.18. Semi-intensive brooding accommodation.

can also affect negative pressure and consequently fan speed. In windy conditions fans controlled to operate at low speed (<30%) can be seen and heard to vary in speed, affecting airflow, creating draughts at floor level and encouraging crowding.

Cycle timers provide more precise control of airflow because fans operating at full speed remain relatively unaffected by wind speed, but always require about 30 s at the beginning of each cycle to develop sufficient negative pressure to encourage airflow. However, providing minimum ventilation in this manner means relatively large volumes of fresh air are drawn at regular intervals through sidewall inlets into the brooding accommodation and, in winter, cold fresh air sinks rapidly to floor level, causing crowding and wet litter. The regular introduction of substantial quantities of cold air in this manner also affects environmental temperature, encouraging brooders to cycle between high and low heat and increasing variation in brooding temperature, adversely affecting bird behaviour.

To avoid these problems, pressure jets can be used to provide an efficient and economical method for mixing fresh cold air with warm internal air and distributing the mixture throughout the brooding accommodation. Unreported trials confirm that this method of ventilation can substantially reduce both litter moisture and heating costs in winter. Pressure jet fan speed must be adjusted to provide sufficient air velocity to prevent temperature stratification. Air speed at floor level should not exceed about 0.2 m/s until chicks are about 6 days of age.

Ceiling or paddle fans can also be used to prevent stratification, but fans should be speed-controlled and rotated anticlockwise in the early days with air directed upwards to prevent down draughts. When chicks are 5 days of age fan rotation should be reversed and speed gradually increased to encourage air movement at floor level.

Lighting

Ducklings are attracted to light and high light intensity will attract birds to a specific area or activity. Illuminating brooders and drinkers attracts day-old ducks to both heat and water. Increasing the light intensity in the relatively cool area surrounding brooders when chicks are about 5 days of age encourages them to explore their environment, exercise and become acclimatized to lower environmental temperature.

Kerosene lamps are used in many parts of the world to provide light and offer the advantage that they can be moved and located as required to assist husbandry during brooding. There are no reported trials describing the effect of light from kerosene lamps on performance, but kerosene lamps certainly provide enough light for people to carry out routine tasks and allow chicks to see, feed and socialize. However, they flicker and this sometimes alarms birds. Pressurized kerosene lamps (Tilley lamps) are a suitable alternative, supplying a very steady light.

Both tungsten and low-energy fluorescent bulbs are widely used to supply artificial light. Tungsten filament bulbs provide the opportunity to regulate light intensity and low-wattage fluorescent bulbs provide low-intensity illumination at very low cost. Providing separate lighting circuits and dimmers for the brooding and the remaining littered or slatted floor areas provides an opportunity to attract ducklings to a particular area and activity. It is essential to provide and programme a timer switch to supply a period of darkness daily to prevent panic in the event of either an interruption in the supply of electricity or equipment failure.

Post-brooding Environment

Ducklings are usually given access to increased floor space or moved to follow-on rearing accommodation between 10 and 17 days of age. In cold climates and during the winter in temperate climates it is prudent to continue to provide artificial heat to maintain environmental temperature above about 17°C until down and feather provide sufficient insulation for birds to tolerate temperatures close to the lower boundary of the thermo-neutral zone of about 10–12°C.

Providing back draught dampers (see Fig. 3.11) and controlling minimum ventilation of fixed-speed fans with an interval timer limits airflow, reduces heat loss and assists in maintaining environmental temperature in cold and cool climates.

Variable-speed fans operating at low power do not generate sufficient air speed and negative pressure to open backdraught shutters, but airflow and heat loss through fans not required for minimum ventilation can be reduced with sliding shutters.

Experience confirms that providing a 'multistat' with a fixed increment of about 1.5°C between stages is more accurate and easier to operate than using several separate thermostats to control temperature in brooding and rearing accommodation ventilated with fixed-speed fans. Thermostats or electronic sensors, used to monitor temperature and initiate ventilation or supply heat when environmental temperature varies outside set limits, must be carefully located to accurately reflect environmental temperature.

Fail-safe ventilation to provide sufficient ventilation to maintain bird comfort and welfare in the event of an interruption in supply, or failure, of electrical components is a legal requirement in many countries. Fail-safe systems are designed so that in the event of power failure or a sudden increase in temperature 'drop out' panels held in place by electromagnets in sidewall and end walls and fan shafts fall open (see Fig. 3.11), providing sufficient 'stack effect' natural ventilation to maintain air quality and prevent a rise in environmental temperature.

Rearing in Hot Climates

In hot and Mediterranean climates birds are frequently reared on slatted flooring because ventilation through the floor assists sensible heat loss. Plastic-coated weldmesh, moulded plastic, perforated metal flooring, wooden slats or bamboo can all be used as flooring materials, but must be carefully constructed and installed because splinters and sharp edges cause scratches, which quickly become infected, causing both sinovitis and septicaemia. Bamboo, in particular, splinters when nailed to supports, causing injury and chronic mortality.

It is essential to supply drainage to prevent a build-up of effluent under slatted flooring and to provide effluent pits with suitable falls to drainage channels and a disposal system. The slatted area, effluent pit and drainage channels should be washed and cleaned regularly to prevent anaerobic digestion of effluent slurry giving rise to foul odours. Alternatively, a scraper system can be used to move effluent to drainage channels and the disposal system.

Traditionally, both breeding stock and table duckling housed semi-intensively have been given access to outside sand runs and straw yards, but increasing concerns about health and hygiene have led to the gradual introduction of concreted yards; however, concrete requires washing frequently to prevent birds soiling feathers and down. It is essential to prevent wild birds and particularly migrating wildfowl from infecting domestic duck with disease. Either fine gauge weld mesh or nylon net can be used to provide a bird-proof enclosure.

In hot climates, providing permanent shade prevents concrete absorbing solar radiation and allows birds to rest on a relatively cool surface away from full sun. In Mediterranean and continental-type climates movable slatted screening provides shade during summer but allows birds to enjoy the benefit of solar radiant heat during the cooler part of the year.

Both breeders and table ducklings are often given access to swimming water located in outside pens. Ducks enjoy bathing and swimming, and both activities assist heat loss because birds enjoy the facility to increase both blood flow and subsequent sensible heat loss from legs, feet and bill, although losing heat in this manner is affected by water temperature. Concave water channels, between 50 and 100 cm wide with a maximum depth of about 15 cm to limit discharge of effluent, should be located in shade to prevent solar radiation from increasing water temperature and to allow birds to avoid full sun when bathing.

Providing running water, even at very slow rates of flow, can reduce water temperature by as much as 10°C, but increasing concerns about pollution and legislation in North America and EC require effluent to be either sprayed over agricultural land, taking care to prevent runoff into drainage ditches and streams, or discharged into an approved effluent disposal system. Alternatively, providing outside pens with an electronically controlled aerosol of water from sprays located about 2 m above floor level can substantially reduce ambient dry bulb temperature without producing effluent. Misting can also be used to moisten the floor surface of outside pens, encouraging latent heat exchange and providing a cool place for birds to rest and socialize during the heat of the day.

Design and Placement of Feeders and Drinkers

Feeders and drinkers should be located to encourage birds to feed, drink and socialize and not simply to facilitate routine husbandry or litter removal. Records of feeding behaviour in a temperate climate (see Figs 4.29 and 4.30) show that ducks eat frequent small meals but require immediate access to drinking water after each meal to clean their bills and assist peristalsis. Frequently drinkers are placed over a slatted area to provide drainage of water splashed during bathing and to prevent litter from becoming wet. However, ducks are naturally attracted to water and locating drinkers over slatted flooring in part litter and slatted floor accommodation often restricts movement and limits effective space (see Fig. 3.19). Ducks enjoy socializing close to water and so distributing feeders and drinkers uniformly over the entire floor area, particularly in hot climates, prevents crowding and encourages effective use of floor space.

Measurements taken between 17 days and slaughter in both Europe and the USA confirm that wastage from conventional feeders is usually at least 4%. Many duck producers use feeders designed for chicken and turkeys, but domestic ducks are not able to peck at food because their bills are adapted with lamellae (comb-like projections along the outer edge of their bill), which are used to filter food particles from water. Feed wastage can occur as birds grasp feed in their bill and then lift their head to swallow it. Figure 3.20 shows a feeder profile designed to prevent wastage. The feeder does not require any adjustment; litter gradually builds up and is about level with the outer edge of the trough when birds are sent to be processed. Trials have confirmed that providing a feeder with this profile will almost eliminate feed wastage and so substantially improve efficiency of feed conversion.

Managing the Environment

Many producers delay providing artificial ventilation until chicks are about 6 days or older to reduce the cost of heating, but ventilating from day-old removes moisture and fumes produced by gas brooders along with products of respiration and excretion, preventing litter becoming moist and sticky and affecting appearance and subsequent growth and development of down. Heat from brooders substantially increases the capacity of air to absorb moisture

Fig. 3.19. Drinker location affecting bird distribution and effective space per bird.

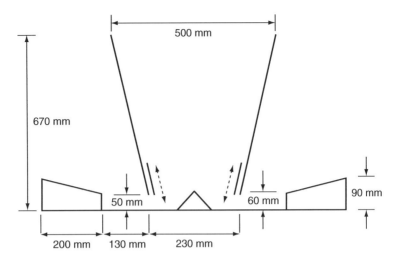

Fig. 3.20. Profile of waste-free feeder.

and so limits the airflow required to maintain a warm and relatively dry environment. However, it is important to regularly increase minimum ventilation to maintain relative humidity below about 85%, which is usually more than sufficient to reduce fumes and gases from combustion to a safe level.

Ducks digest food and excrete liquid faeces as 'squitter'. The liquid effluent is absorbed into warm and dry litter encouraging growth of microorganisms and the release of ammonia into the brooding environment. It is essential to spread fresh clean litter daily over the brooding floor area and to increase minimum ventilation every other day to prevent ammonia (measured before bedding with fresh litter) exceeding about 10–12 ppm. Exposure to ammonia levels above about 25 ppm for several hours can cause blindness, as well as affecting general health and physical performance. The moisture content of litter increases daily from about 5 days and this encourages growth of microorganisms. By about 10 days the airflow required to maintain ammonia below 12 ppm is normally more than sufficient to prevent relative humidity exceeding 85%.

Ambient temperature and ventilation affect brooder radiant temperature because in cold conditions airflow reduces environmental temperature, and the control system responds by increasing gas pressure and subsequent brooder radiant temperature. Environmental temperature then increases towards and then beyond the 'set point' and the control system reduces gas pressure and temperature declines, completing the cycle. Ambient temperature, minimum airflow, air speed, location of sensors, along with sensor differential (difference in temperature between heat on and off to maintain a controlled set-point temperature) all affect the nature and duration of the cycle and consequent chick comfort and behaviour. Farm staff should be encouraged to observe and become familiar with the effects of these factors on the control cycle and consequent behaviour along with the effects of adjusting temperature, airflow and speed on air quality and litter condition and behaviour (appearance, activity and noise).

Ducklings brooded in cold and cool temperate climates are frequently given more space or moved to adjacent accommodation (depending on ambient temperature) between about 12 and 17 days of age. Increasing space and providing clean dry bedding reduces both humidity and ammonia, but it is essential to maintain air quality and provide sufficient artificial heat in cold weather to prevent crowding and the litter from becoming wet and dirty. Many producers economise by substantially reducing heat as early as 10 days, but replicated commercial trials confirm that reducing environmental temperature in this manner causes crowding and reduces feed intake and subsequent live weight at slaughter.

Total feed intake per day, total body weight, number of birds placed and similar parameters are widely used to calculate ventilation requirements for broiler chicken. Attempts to use these parameters to calculate and supply minimum ventilation for successive batches of ducklings, brooded and reared in different types of accommodation at several locations over about 12 months, were confounded by different systems of ventilation and similar factors. However, analysis of recorded minimum ventilation settings confirmed that the minimum airflow (m^3/s) needed to maintain air quality (humidity <85%, ammonia <15 ppm and carbon dioxide <0.3%) for any particular location was consistent and provided a reliable guide to minimum ventilation required between day-old and slaughter at that location.

Rules for setting maximum and minimum ventilation based on feed intake (m^3/s per tonne of feed eaten per day) or body weight (m^3/s $kg^{0.75}$) are all very well, but observation of bird behaviour and litter condition are the keys to achieving optimum performance.

References

Charles, D.R., Elson, H.A. and Haywood, M.P.S. (1994a) Poultry housing. In: Wathes, C.M. and Charles, D.R. (eds) *Livestock Housing*. CAB International, Wallingford, UK, pp. 266.

Charles, D.R., Elson, H.A. and Haywood, M.P.S. (1994b) Poultry housing. In: Wathes, C.M. and Charles, D.R. (eds) *Livestock Housing*. CAB International, Wallingford, UK, pp. 265.

Charles, D.R., Elson, H.A. and Haywood, M.P.S. (1994c) Poultry housing. In: Wathes, C.M. and Charles, D.R. (eds) *Livestock Housing*. CAB International, Wallingford, UK, pp. 251.

Edwards, P. (1985) Duck/fish integrated farming systems. In: Farrell, D.J. and Stapleton, P. (eds) *Proceedings of a Workshop, Duck Production Science and World Practice*. University of New England, Armidale, NSW, Australia, pp. 267–291.

Hunter, J.M. and Scholes, J.C. (1954) *Profitable Duck Management*, 9th edn. The Beacon Milling Company, Cayuga, New York.

Lee, A.R. (1915) *Duck Raising*. US Department of Agriculture, Farmers Bulletin No 697, US Government Printing Office, Washington, DC.

Mount, L.E. (1979) *Adaptation to Thermal Environment*. Edward Arnold, London.

Scott, M.L. and Dean, W.F. (1991) *Nutrition and Management of Ducks*. M.L. Scott of Ithaca, Ithaca, New York.

4 Husbandry of Table Duckling

This chapter is chiefly concerned with the growth and development of the Pekin duck, which is by far the most widely used commercial breed of domestic duck. It has been subject to intense selection since its importation into the USA and Europe in the late 19th century and Fig. 4.1 illustrates the effects of that selection on its growth. As Fig. 4.1 shows, there have been substantial changes in growth rate over the years, but Fig. 4.2 makes the point that the pattern of growth, relative to adult weight, has changed very little. All genotypes achieve mature weight at about 12 weeks, by which time maturation of the primary and secondary wing feathers is complete (see Fig. 6.3).

The reason for this rapid growth to mature weight, compared with other domestic poultry, is that wild Mallard from which nearly all breeds of domestic ducks are derived, fledge by approximately 50–55 days (Queeny, 1983; Nielson, 1992) and migrate at around 12–16 weeks of age. Growth of bone, muscle and feather is therefore genetically controlled and coordinated to achieve critical ratios between live weight, breast muscle mass and wing area by that age. Data will be presented in Chapters 5 and 9 (this volume) to show that growth and body composition of Pekin ducks are still profoundly affected by relationships and constraints inherited from wild Mallard.

Feathering

Table 4.1 describes the sequence of growth of feather and down in Pekin duck, which is affected by genetic factors inherited from its wild forebear the Mallard. Ducks hatch covered with a layer of down that provides good thermal insulation. When coated with oil from the feathers of the brooding duck, down is sufficiently waterproof to allow birds to swim soon after hatching. Down grows rapidly and, by about 10 days, provides sufficient insulation to enable young duck to survive in temperate climates without maternal brooding. From about

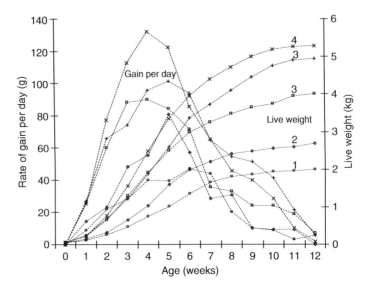

Fig. 4.1. Growth and rate of gain to mature weight of Pekin. (From Gonzalez and Marta, 1980; Testik, 1988; Cherry, 1993; Maruyama *et al.*, 2001.)

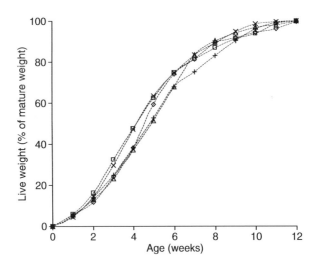

Fig. 4.2. Relationship between age and growth expressed as a percentage of mature weight. Data and key as in Fig. 4.1.

14 days the structure of down changes and becomes increasingly filamentous until, by 30 days, growth of juvenile down is complete with fully formed down clusters providing sufficient thermal insulation to enable the growing bird to maintain body temperature at low ambient temperatures.

Feather growth over the breast area begins at about 14 days and is complete by about 35 days of age. Feathers on the back start growing rapidly from about

Table 4.1. Relationship between age and relative development of feather and down of Pekin duck.

Age (days)	0	4	8	12	16	20	24	28	32	36	40	44	48	52	56
Down	X	X	X	X	X	X	X	X							
Breast feather						X	X	X	X	X	X				
Primary feather							X	X	X	X	X	X	X	X	X
Back feather										X	X	X	X	X	X

30 days and the back and flanks are fully feathered by about 50–60 days. Primary and secondary wing feathers can be measured from as early as 24 days and grow rapidly to maturity by about 56 days (see Fig. 6.3). This is about the age when juvenile wild Mallard fledge. At about 54 days, juvenile breast feather is replaced. This moult, sometimes referred to as 'second feather', starts at the upper neck and the relatively large and rapidly developing feather follicles that appear are often described as 'pin feathers'. These are difficult to remove during plucking, which affects the subsequent appearance of the processed bird, thus limiting the opportunity to market birds until this moult is complete at about 70 days.

Duck feather and, in particular, down are valuable by-products used as insulation in clothing and duvets and for filling cushions in high-value furniture. Feather and down recovered after plucking are washed, dried and then separated into feather and down using specialized equipment. The weight of washed and dried feather and down is usually about 3.5% of the live weight. Data are presented in Chapters 3 and 5 (this volume) showing that growth of feather is affected by both genotype and nutrition.

Husbandry and environment can also affect both growth and quality of feather and down. In particular, crowding and wet litter affect the normal growth and development of down clusters because down filaments (the fine strands that grow to make a down cluster) become wet and dirty and stick together. This significantly reduces quality, measured in terms of filling power, down cluster size and economic value. It is important to provide sufficient heat during brooding for birds to remain comfortable without crowding and to supply clean litter and fresh air. About $0.2\,m^2$ of space per bird is needed to encourage good growth and development of feather and down, which helps prevent birds from inadvertently scratching one another when resting and during transportation for processing.

Temperature, Feed Intake and Growth

Figures 4.3 and 4.4 describe the effect on growth and feed consumption of rearing Pekin at a relatively stable post-brooding temperature in several well-replicated trials. Analysis of these data provides the following estimates of the effect of temperature on live weight and feed intake from 21 to 48 days.

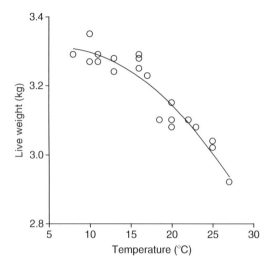

Fig. 4.3. Relationship between temperature and growth to 48 days recorded in five trials for Pekin ducks reared from 21 to 48 days at a constant temperature. (From Cherry, 1993.)

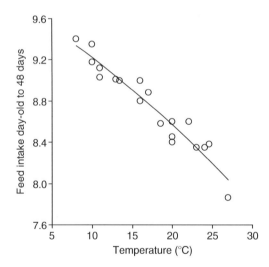

Fig. 4.4. Relationship between temperature and feed intake recorded in five trials for Pekin ducks reared from 21 to 48 days at a constant temperature.

Live weight (kg) at 48 days $= 3.281 + 0.0102x - 0.0009x^2$

Feed consumption (kg) to 48 days $= 9.751 - 0.0466x - 0.0006x^2$

where $x =$ temperature (°C; mean of daily temperatures).
        In another set of six trials, live weight and feed intake were recorded weekly from 28 to 35 days and from 35 to 42 days of age. Figures 4.5 and 4.6 show

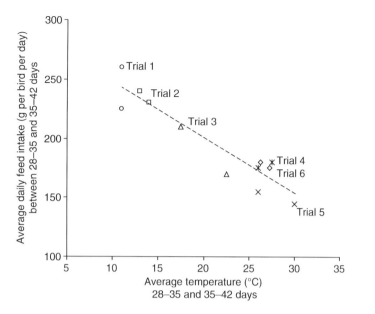

Fig. 4.5. Relationship between average temperature (average of daily maximum and minimum temperature) and daily feed intake between 28–35 days and 35–42 days recorded in six trials for Pekin given diets of similar nutrient composition.

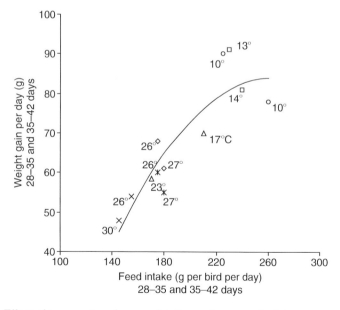

Fig. 4.6. Effect of temperature (mean of daily maximum and minimum temperatures) on feed intake and rate of gain between 28–35 and 35–42 days recorded in six trials where Pekin of the same genotype were given feed of similar nutrient content. Data and key as in Fig. 4.5.

the effect of temperature on daily feed intake and on subsequent rate of gain in these trials. Analysis of the data provides estimates of the effect of temperature between 28 and 42 days as follows:

Daily feed intake (g) = 295.21 − 4.7521x

where x = mean of daily maximum and minimum temperatures.

Daily rate of gain (g) = −110.52 + 1.4811x − 0.0028x²

where x = daily feed intake (g).

Figure 4.7 further describes the effect of daily temperature (mean of maximum and minimum temperature) on feed intake recorded between 31 and 48 days. Detailed analysis of feed intake at regular intervals of 6 h in this trial shows that when daytime temperature increased beyond about 22°C birds preferred eating between dusk and dawn, and when midday temperatures exceeded about 26°C birds consumed as much as 80% of their daily total feed intake between midnight and 06.00 h.

The results of the trials summarized in Figs 4.3–4.7 support the conclusion of Emmans and Charles (1977) that rate of feed intake for a given feed by a particular genotype grown under defined husbandry conditions will depend on the temperature of the environment in which the bird is kept. The growing duck attempts to achieve its genetic potential growth, but actual rate of gain per day is governed by physiological age (see Fig. 4.1), nutrient intake (see Chapter 5, this volume) and the extent to which the bird is able to dissipate metabolic heat. Birds, except in the short term, are unable to store heat, so heat loss must equal rate of heat production for the bird to stay in equilibrium. Feed intake is therefore controlled by the bird's ability to dissipate heat arising from metabolic activity and this is governed by effective temperature.

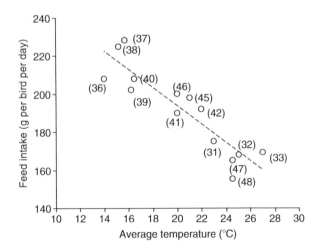

Fig. 4.7. Effect of daily temperature (mean of maximum and minimum temperatures) upon daily feed intake of Pekin between 31 and 48 days with age recorded in brackets.

High Temperature and Performance

Figure 4.8 shows the effect of average post-brooding temperature, recorded in a climate with temperate winters and hot humid summers, on feed intake and subsequent live weight recorded at 35, 42 and 48 days. Analysis of the results of these trials provides the following estimates for the effect of temperature (x, °C) on 48-day live weight and feed intake to 48 days.

Live weight (kg) = $3.557 + 0.007x - 0.0016x^2$

Feed intake (kg) = $8.481 - 0.021x - 0.0025x^2$

Figures 4.9 and 4.10 compare the effect of temperature on feed intake and live weight recorded in these trials with the trial data, recorded in the UK, described in Figs 4.3 and 4.4. Rearing in open-sided and naturally ventilated accommodation in a relatively humid climate substantially reduced both feed intake and growth to 48 days compared to birds of a similar genotype reared in the UK trials where gas heating was used to maintain a relatively constant post-brooding temperature. The artificial heating reduced relative humidity and increased the opportunity for birds to lose heat when panting relative to birds exposed to natural summer temperatures in a humid climate. Feed intake and subsequent growth is governed by the bird's ability to lose heat arising from metabolic activity and, as dry bulb temperature increases, the proportion of metabolic heat lost directly declines and indirect loss by evaporative cooling during respiration increases; but this is markedly affected by atmospheric humidity.

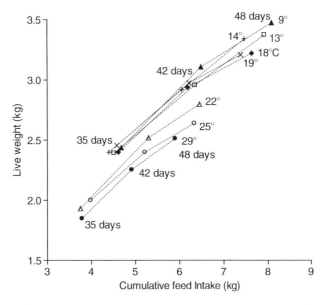

Fig. 4.8. The effect of age and post-brooding environmental temperature (mean of daily maxima and minima) on live weight for successive groups of Pekin of the same genotype reared in the same accommodation and given feed of the same nutrient composition.

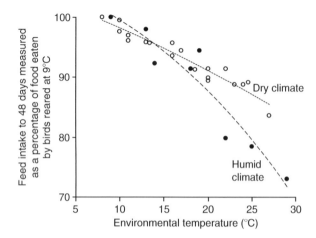

Fig. 4.9. Effect of temperature on feed intake to 48 days for Pekin reared in different post-brooding climates given diets of similar nutrient composition. Pekin in the temperate climate were reared intensively on litter and slats using indirect gas heating to maintain temperature. A similar genotype was reared in a climate with temperate winters and hot, humid summers, on litter in open-sided accommodation with insulated roofing.

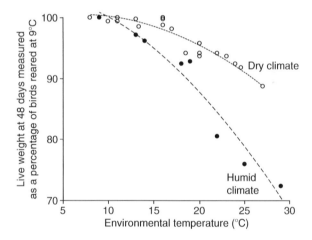

Fig. 4.10. Effect of temperature on growth for Pekin reared in different climates. Genotype, nutrition and accommodation as in Fig. 4.9.

Both evaporative cooling of the house and misting birds at frequent intervals with a fine aerosol of water are widely used in the broiler industry in southern USA to reduce heat stress and improve performance. Large-scale trials confirm that evaporative cooling was effective in reducing environmental dry bulb temperature for birds housed intensively on all litter. Figure 4.11 shows that environmental temperature declined in a curvilinear manner with increasing ambient temperature, because increasing ambient temperature reduced ambient water vapour pressure and increased both evaporation and sensible heat loss of air drawn through the evaporative coolers. Figure 4.12 shows that evaporative cooling can reduce environmental temperature to within about 2°C of ambient wet bulb temperature.

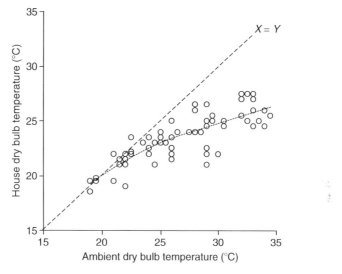

Fig. 4.11. The relationship between ambient and environmental dry bulb temperature for Pekin duck housed intensively on all litter. Birds were provided with evaporative cooling with air drawn through inlets made of corrugated cellulose kept wet with water. Temperature was recorded with a whirling hygrometer twice daily at random times between 18 and 48 days.

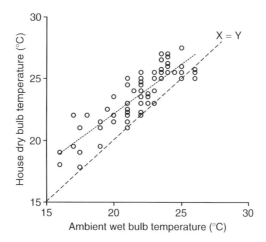

Fig. 4.12. The relationship between ambient wet bulb temperature and house dry bulb temperature for accommodation provided with evaporative cooling. Accommodation and temperature recorded as described in Fig. 4.11.

Relative humidity can also be used to measure and describe the moisture content of air and Fig. 4.13 shows the effect of ambient relative humidity on the effectiveness of evaporative cooling. Cooling reduced average temperature (mean of daily maximum and minimum dry bulb temperatures) by about 4°C, but reducing dry bulb temperature in this manner increased growth in replicated trials to 48 days by only 3%. This was substantially less than would be expected from UK trials investigating the effect of temperature on growth, as described in Fig. 4.3. This was because reducing dry bulb temperature by evaporative cooling was achieved by increasing humidity, and this substantially reduced the opportunity for birds to lose heat indirectly by evaporation of moisture through increased respiration. Losing heat in this manner depends not on the difference between ambient and environmental dry bulb temperature but on differences in water vapour pressure between the bird and its environment (see Fig. 4.10). At high environmental temperature the difference in water vapour pressure between the bird and its environment is more important than the dry bulb temperature.

Misting can be used to reduce heat stress for ducks grown at high temperature, but to be effective it is essential to provide a uniform distribution of water aerosol over the birds. This is difficult to provide when birds are housed on litter in naturally ventilated and open-sided accommodation. However, trials placing misting nozzles in front of large-diameter (60 cm) fans, located at 15 m intervals along the length of the building, were successful in achieving a reasonably even distribution of a water aerosol and improved weight for age in replicated commercial trials by about 3%. However, misting birds in this manner adversely affected both litter condition and feathering.

Tunnel ventilation, high-speed pressure jets and ceiling fans are all widely used in hot climates to reduce heat stress. Extensive commercial trials with both

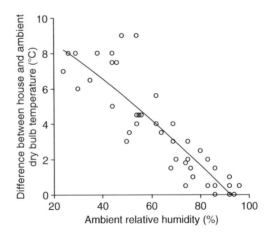

Fig. 4.13. The relationship between ambient relative humidity (%) and reduction in house dry bulb temperature for ducks provided with evaporative cooling. Accommodation and temperature recorded as described in Fig. 4.11.

tunnel ventilation and pressure jets confirm that growing ducks prefer to rest in a stream of fast-moving air, but ventilating birds in this manner had no measurable effect on their feed intake or growth. A common problem was that birds tended to congregate in the area of highest air speed, thus reducing effective space per bird and affecting both litter quality and feathering.

Increasing the age at which birds are processed is widely used by producers to offset the effects of high temperature on growth during summertime. However, Fig. 4.1 shows that daily rate of gain declines rapidly after 35 days and increasing age reduces the efficiency of feed conversion (see Fig. 4.14).

A better method of increasing weight for age is to use a faster-growing genotype. Figure 4.14 describes the effect of age on growth of selected and unselected Pekin genotypes reared at high temperature (24°C) and shows that while feed conversion was similar to 56 days, the faster-growing genotype was substantially more efficient to any given weight. However, increasing weight for age and reducing killing age would require a genotype selected for maturity and body composition as well as growth (see Chapter 9, this volume).

Another way to increase live weight yield per square metre is to kill the two sexes at different ages. Figure 4.15 shows the substantial effect of sex on relative growth in two genotypes between 40 and 52 days of age. Figure 4.16 shows a substantial difference between genotypes when growth of females is expressed as a percentage of male live weight, although relative growth declined by about 0.35% per day between 40 and 52 days for both genotypes. Evidence will be presented in Chapter 5 (this volume) to show that females are earlier maturing than males in terms of body composition and can be processed at least 2–3 days earlier than males.

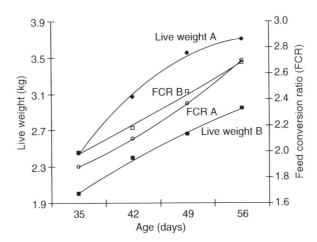

Fig. 4.14. Relationship between age, growth and efficiency of feed conversion ratio (FCR) for genotypes A and B reared at about 24°C (measured as an average of daily minimum and maximum temperatures) on all litter in open-sided naturally ventilated accommodation and given diets of the same nutrient concentration.

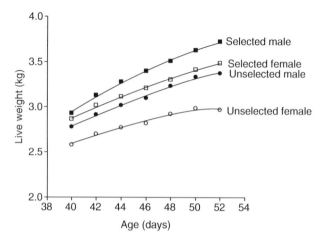

Fig. 4.15. Rate of growth between 40 and 52 days for selected and unselected Pekin genotypes grown at high ambient temperature. Birds identified to genotype and sex were reared in open-sided accommodation and weighed at 2-day intervals.

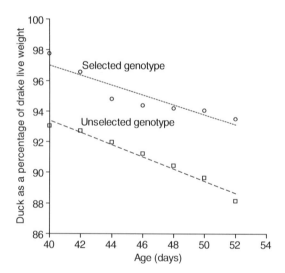

Fig. 4.16. Relationship between age and live weight of females from selected and unselected genotypes expressed as a percentage of male weight. Data as in Fig. 4.15.

Low Temperature and Performance

Ducks can tolerate low temperatures because they are covered with down, feather and a layer of subcutaneous fat to provide insulation, and they employ countercurrent heat exchange and control blood supply to reduce heat loss from the legs and feet when swimming or standing in cold water. There are no reports in the literature of the effect of very low temperatures on performance, but experience in Eastern Europe shows that, where birds are given access to bathing water in outside pens during daylight hours, they bathe, swim and socialize and appear comfortable when ambient temperatures are close to freezing.

Figure 4.17 describes the results of a trial to investigate the effect of temperature in the range 8–15°C on growth to 48 days, and shows that reducing post-brooding temperature increased feed intake almost linearly, but had no substantial effect on growth. However, growth and feed intake in this trial may have also been affected by reducing post-brooding temperature at a relatively early age, because Fig. 4.18 shows that ducks brooded at several commercial locations in summer achieved a slightly greater daily rate of gain to 21 days compared to birds of the same genotype brooded at the same locations in winter. Figure 4.19 shows the importance of brooding and post-brooding environment on growth to both 21 and 42 days for birds reared in either winter or summer.

Experience in cold climates with growing ducks given access during daylight to outside pens shows they can tolerate temperatures as low as 2–4°C from as early as 24 days, but when relative humidity exceeds about 75% for any length of time litter becomes wet and dirty, affecting feathering, growth and bird health. Figure 4.20 describes the relationship between age, daily average temperature and humidity during winter for a large flock of growing ducks reared intensively on all litter, and Fig. 4.21 shows the effect of diurnal variation in temperature on relative humidity for the same flock of birds at 32 days. The data in both Figs 4.20 and 4.21 show that when environmental temperature fell below about 8°C, relative humidity increased beyond 80%, which rapidly causes litter to become saturated with moisture, affecting growth, health and feathering and subsequent processed quality and economic performance.

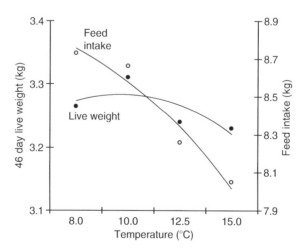

Fig. 4.17. Effect of low temperature on growth and feed intake of Pekin to 46 days. Birds were housed intensively on litter and post-brooding temperature was gradually reduced to provide 8°C, 10°C, 12.5°C and 15°C (mean of daily maximum and minimum dry bulb temperatures) from 18 to 46 days of age.

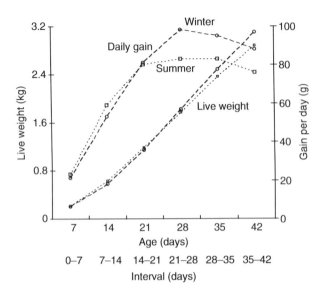

Fig. 4.18. Effect of rearing commercial flocks of Pekin in winter and summer on growth and rate of gain to 42 days. Birds of the same genotype were reared at the same locations and given feed of the same nutrient content in winter and summer when average ambient temperatures (mean of daily maximum and minimum temperatures) were 7°C and 19°C, respectively.

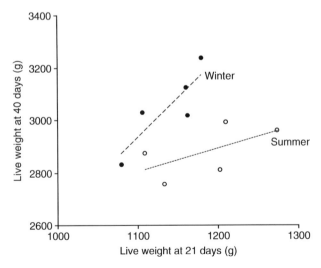

Fig. 4.19. Relationship between live weight at 21 and 42 days for Pekin reared at the same locations during summer and winter. Genotype, accommodation and ambient temperatures as in Fig. 4.18.

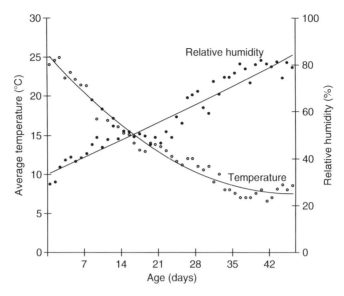

Fig. 4.20. The relationship between daily average temperature and relative humidity (mean of data recorded at 30 min intervals) for Pekin between day-old and 48 days housed intensively on straw litter and placed at 4.5 birds per m².

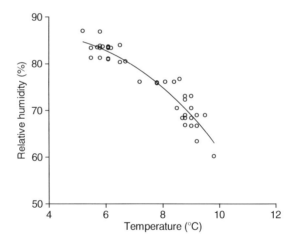

Fig. 4.21. Relationship between temperature and humidity recorded at 30 min intervals over 24 h for Pekin aged 32 days. Average temperature and relative humidity were 7.3°C and 77%, respectively. Accommodation and space allowance as in Fig. 4.20.

It is important to provide birds with clean and dry bedding by spreading fresh litter over the entire littered area each day, and to encourage litter fermentation by maintaining environmental temperature above 8°C by providing insulated accommodation, equipped with windproof air inlets and fan outlets (see Chapter 3, this volume) along with precise control of minimum ventilation. However, in cold climates where night-time ambient temperatures regularly fall well below freezing, some form of supplemental heat may be required or it may be worth considering placing more birds per square metre to increase both metabolic heat and composted deep litter temperature.

Temperature and Behaviour

In Eastern Europe, where ambient temperature in winter regularly falls well below freezing and the house temperature at night occasionally falls below 5°C, fully feathered birds usually rest in large groups (rafts) on litter to reduce metabolic heat loss and take advantage of composted litter temperature. Birds rest with legs and feet drawn under their body to reduce both radiant and convected heat loss and commonly place their head under a wing to reduce heat loss from the bill. Feeding, drinking and social activities are restricted and birds are unusually quiet.

As temperature increases towards 12°C birds rest on litter in rafts frequently with feet and legs exposed. The birds appear comfortable and can be seen preening, and as temperature increases beyond about 15°C there is a general increase in social activity.

Above 18°C birds frequently rest on any available wire or slatted floor area, raft size decreases with increasing temperature and birds rest with legs and feet fully exposed in an effort to achieve maximum sensible heat loss. When resting in rafts, individual birds can occasionally be seen panting to lose heat via respiration. Birds spend increasing time drinking, preening and enjoying social activities.

Above about 22°C birds start panting intermittently, increasing their rate of respiration and evaporating moisture from the trachea to lose heat and maintain homeostasis. As temperature increases above 26°C, particularly when relative humidity exceeds 60%, panting becomes very frequent, with some birds panting by 'gular flutter', which increases evaporative heat loss and convective heat loss by forced convection.

When the temperature exceeds 32°C, birds become increasingly lethargic and reluctant to move, with no social or feeding activity and little 'duck noise'. Above 35°C, birds appear increasingly distressed, sitting with neck and head extended panting by gular flutter and water can be seen dripping from the bird's bill.

In the tropics where ambient temperature can frequently exceed 35°C it is sensible to supply bathing water because the birds can increase both blood flow and sensible heat loss from the bill, legs and feet when swimming and bathing. One reason why ducks are able to tolerate high temperature is that they have a small oesophagus with no crop and so they have only limited ability to store feed. This reduces heat arising from digestion and subsequent metabolic activity, preventing hyperthermia when daytime temperature increases.

The terms zone of thermal neutrality, thermal comfort and optimal productivity (Mount, 1979) measured in terms of dry bulb temperature are widely used to describe the relationship between temperature and metabolic activity, behaviour and welfare. However, the evidence from reported trials shows that factors such as age, behaviour, water vapour pressure, air speed and litter temperature and condition can all affect the bird's response to temperature. Evidence will also be presented later in this chapter to show that space per bird, along with minimum ventilation required to provide oxygen and remove carbon dioxide and ammonia, and similar factors should also be taken into account when describing the effect of temperature on behaviour and performance.

Space and Performance

Producers in Long Island, New York (see Chapter 3, this volume), until about 1970 used a system that provided growing ducks with increased space as they grew. Several ages of birds, separated by fencing, were accommodated in the same building and each week the oldest group was sent for slaughter and another batch of day-old ducks was placed. A similar method for increasing productivity is still used in Europe. Ducks are brooded in about a third of the house and moved into the remainder of the accommodation when artificial heat is no longer required. The brooding area is then cleaned and another batch of ducks is placed, brooded and moved in turn. Both systems increase productivity measured in terms of kilogram of duck produced per square metre per annum, but rearing birds of different ages in close proximity in this manner can affect health, and increase mortality and rejection during processing due to sacculitis and salpingitis.

In large operations, table ducklings are now almost always reared as single-age groups in one house and often the 'all in, all out' principle is applied to a whole site, to achieve the stringent health and hygiene standards required by state and retail market outlets. This leads to the question of what stocking rate should be used when it is to remain constant from day-old until slaughter.

There are no reports in the literature of the effects of space allowance on physical performance of table duckling. However, results from an unreported trial described in Figs 4.22 and 4.23 show that space allowance during brooding to 16 days of age has a small, but measurable, effect on both feed intake and growth to 45 days for birds subsequently given the same space allowance from 16 to 45 days of age. Results of another large and well-replicated trial described in Fig. 4.24 show that increasing the number of birds placed per square metre at 17 days reduced growth to 46 days, and analysis of the data provides the following estimates of the effect of space in the range 4–8 birds per m^2 on growth and cumulative feed intake to 46 days, with live weight and feed intake expressed as a percentage of live weight and cumulative feed intake of birds placed at 4.1 birds per m^2.

Live weight (%) = $97.253 + 2.123x - 0.348x^2$

Cumulative feed intake (%) = $114.23 - 3.512x$

where x = number placed per square metre at 17 days of age.

Space allowance had no significant effect on either mortality or rejection during processing, but increasing the number of birds beyond about 6 birds per m^2 adversely affected feathering and, occasionally, grading at processing.

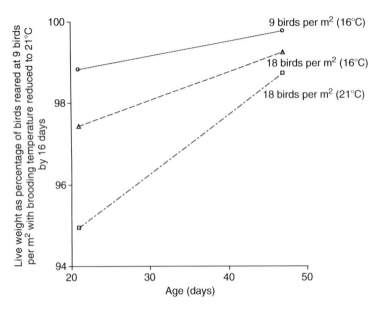

Fig. 4.22. Effect of space allowance and brooding environmental temperature upon growth to 47 days, measured as a percentage of birds reared at 9 birds per m² and 21°C to 16 days of age. Birds were reared at either 9 or 18 birds per m² to 16 days and environmental temperature was gradually reduced to provide either 16°C or 21°C by 16 days. From 16 days space allowance for all birds was increased to 7 birds per m² and birds were given the same environmental temperature to 45 days.

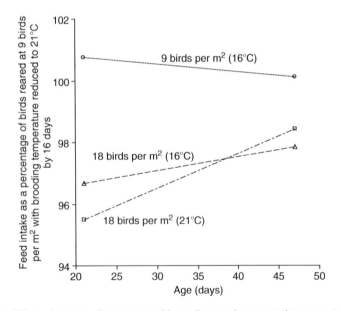

Fig. 4.23. Effect of space allowance and brooding environmental temperature upon food intake to 45 days, measured as a percentage of birds reared at 9 birds per m² with environmental temperature, reduced to 21°C by 16 days of age. Space allowance and environmental temperature as in Fig. 4.22.

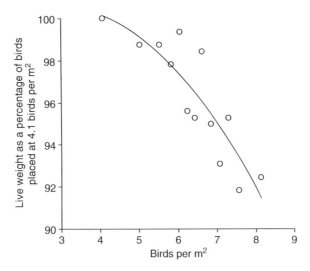

Fig. 4.24. The relationship between space allowance and growth to 46 days measured as a percentage of birds placed at 4.1 birds per m². Birds supplied from the same breeding flocks and given the same space allowance during brooding were placed at weekly intervals when 17 days old into intensive accommodation between 4 and 8 birds per m². There were 13 space treatments replicated four times at regular intervals over one calendar year; birds were given feed of similar nutrient concentration.

Figures 4.25 and 4.26 describe the effect of space allowance on feed intake and growth of Pekin ducks housed intensively on litter and slats to 45 days of age. They show that increasing the number placed per square metre at 17 days significantly reduced both cumulative feed intake and live weight to 45 days by about 3% and 4%, respectively. It is important to note that Figs 4.25 and 4.26 also show that increasing the number of birds per square metre reduced growth and feed intake from an early age, before space might reasonably be considered to limit activity or affect behaviour.

Increasing the number placed per square metre in a given house will increase group size and litter temperature and also reduce feeding and drinking space per bird. Analysis of two well-replicated commercial trials concluded that group size in the range 90–1100 birds placed at either 5.7 or 8.3 birds per m² had no significant effect on either feed intake or growth to 45 days of age. Similarly, increasing feeding and drinking space either separately or concurrently had no significant effect on either feed intake or growth of birds stocked at either 5.7 or 8.3 birds per m². Measurement of litter temperature confirmed that increasing the number of birds per square metre substantially increased bird effluent, litter fermentation and subsequent litter temperature by about 3–4°C, but reducing space allowance reduced feed intake before fermentation was sufficiently advanced to increase litter temperature. Increasing the number placed per square metre reduced both

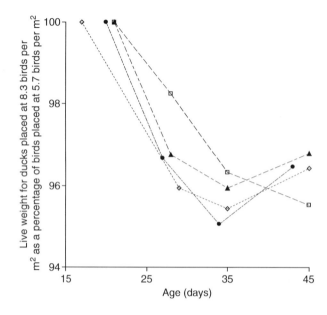

Fig. 4.25. Effect of space allowance on growth to 45 days recorded in four separate trials. Birds of the same genotype were reared together to 17 days and then placed at either 270 or 390 birds per pen providing 5.7 or 8.3 birds per m² and there were 12 replicates of each treatment in each trial. Birds were housed in intensive accommodation on litter and slats, and live weight and feed intake were recorded from between 17 and 21 days at intervals to 45 days of age.

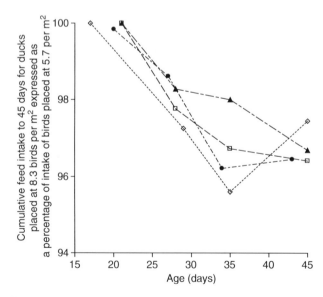

Fig. 4.26. Effect of placing birds at 8 birds per m² on cumulative feed intake to 45 days. Feed intake was recorded from between 17 and 21 days at intervals to 45 days of age. Trial design and layout as in Fig. 4.25.

live weight (see Fig. 4.24) and weight of feather, but had no substantial effect on feather yield when expressed as a percentage of live weight. Increasing the number of birds placed per square metre had no significant effect on either mortality or the rejection rate during processing in any of the trials reported.

Figure 4.27 summarizes the results of eight stocking density trials. There were substantial differences between trials in weight for age arising from differences in genotype and age at slaughter, but all trials show a common slope, with final live weight declining by between 35 and 40 g (about 1% of final live weight) over the range 5–8 birds per m². Figure 4.25 confirms that the effect of space allowance on performance is reasonably predictable for a given genotype reared in similar accommodation and given feed of the same nutrient composition.

Figure 4.28 shows that increasing the number of birds grown per square metre of floor space has a substantial effect on productivity measured as kilogram of duck produced per square metre of floor space, reducing capital requirements for housing and fixed costs per kilogram of duck produced. However, increasing the number placed per square metre reduces both feed intake and weight for age (see Figs 4.25 and 4.26), and at a stocking density of 8.3 birds per m² it would be necessary to increase age at slaughter by about 2 days to achieve a weight similar to that of birds placed at 5.7 birds per m², thus increasing cumulative feed intake and affecting economic performance.

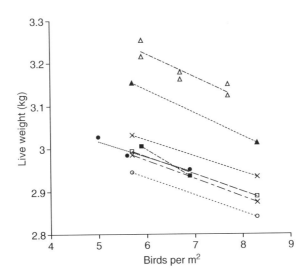

Fig. 4.27. Relationship between number placed per square metre of available floor space and growth of Pekin recorded in eight separate trials. Different genotypes were used in the trials. Birds in each trial were brooded together to about 17 days when space allowance was adjusted. Reported live weight was recorded between 40 and 45 days of age. Feed of similar nutrient composition was provided to birds in all trials.

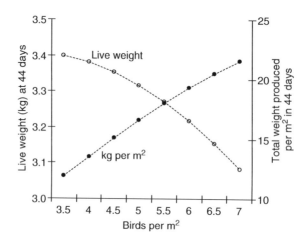

Fig. 4.28. Relationship between number of birds placed per square metre, growth and weight of duck produced per square metre to 44 days of age based on data shown in Fig. 4.24.

The number placed per square metre can profoundly affect financial performance; marginal cost–benefit analysis provides a suitable method to determine the optimum economic space allowance for a given genotype and environment, taking into account the effects of space allowance on total live weight produced per square metre, feed intake and growth, along with age at slaughter required to achieve a particular live weight. However, in many countries it is necessary to take account of bird welfare constraints designed to prevent stress and allow birds reasonable standards of comfort. Welfare constraints vary from country to country, but usually provide guidelines or limits on the number of birds or weight of birds produced per square metre of accommodation.

Light and Performance

Mallard migrate to northern latitudes to breed in the springtime where their progeny may experience a daylength exceeding 20 h from hatching to somatic maturity. Most species of wildfowl do not have a crop to store feed, but long days provide an opportunity for birds to feed frequently and achieve sufficient growth to participate in the annual migration south in autumn.

Domestic ducks, like their wild forebear the Mallard, do not have a crop and Figs 4.29 and 4.30 show that Pekin fed *ad libitum* prefer to eat small amounts at frequent intervals, although feeding behaviour and intake are affected by genotype, environmental temperature and daylength. At night birds rest in groups or rafts, seeking security and comfort through social contact and, when moving about, occasionally scratch one another on the back and flanks. These vulnerable areas are covered with only filamentous down, until protected by body feathers from about 40 days (see Table 4.1).

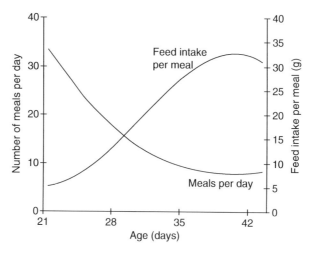

Fig. 4.29. The relationship between age and the number of meals and feed intake per meal for Pekin with a mature live weight of 3.6 kg maintained at about 12°C given an 11.8 MJ/kg and 18% protein diet.

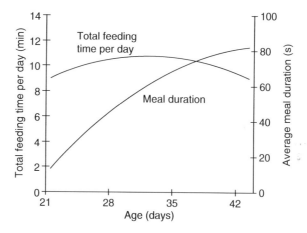

Fig. 4.30. The relationship between age, average meal duration and total time spent feeding per day.

Trials with intermittent light in windowless houses have shown that a programme of 1 h light and 2 h darkness and other similar combinations of light and darkness reduced both feed intake and growth. It was necessary to increase age at slaughter by about 1 day to achieve the same live weight as birds given 23 h of light and 1 h of darkness. Commercial-scale replicated trials also confirmed that intermittent light substantially increased the proportion of birds downgraded for back scratching and bruising, and reduced water consumption by about 30% and subsequent feather yield by about 7% relative to birds given 23 h light and 1 h of darkness.

Oil, gas or electricity can be used to provide light. It is customary to provide about 10 lx at day-old, gradually reducing light intensity to between 1 and 2 lx for birds reared in intensive accommodation. Light intensity greater than about 5 lx encourages feather pulling. Ducks have good vision at low light intensity and moonlight can often provide sufficient light for birds in windowed accommodation to feed, drink and socialize.

Fencing or pliable metal or plastic surrounds are widely used to keep ducks close to the brooder. This is moved outwards as birds grow. However, light is equally effective in limiting and encouraging the movement of chicks. At day-old, light is limited to the brooder area but at about 4 days the lit area is gradually increased to include the remaining floor area to encourage birds to explore their surroundings.

A short period of darkness should be given each day to accustom birds to darkness and so prevent panic in the event of any interruption in power supply. Staff should have the facility to temporarily increase light intensity for routine husbandry and inspections.

Brooding Environment

Domestic ducks in common with other domestic poultry are unable to control body temperature for several days after hatching. In temperate climates it is necessary to supply some form of artificial heat until down cover provides sufficient insulation to enable birds to maintain body temperature between 41.2°C and 42.2°C (Mount, 1979).

Figure 4.31 describes the relationship between age and brooder temperature for table duckling housed intensively at about 15 birds per m² when located in a cool temperate climate. Figure 4.32 shows that brooder temperature declines in a curvilinear manner with distance, giving birds an opportunity to select a comfortable temperature for rest and sleep. However, it is important to avoid placing too many ducks under brooders to prevent birds scrambling over one another to reach a warm place to rest. Competition also means that, when birds finally achieve a comfortable place to rest, they become reluctant to move, affecting feed intake and subsequent growth.

Litter such as sawdust, shavings, straw and rice hulls usually contains more than 16% moisture on delivery. It should be spread and dried with artificial heat prior to the arrival of day-olds to provide chicks with a warm and dry brooding environment. It is necessary to adjust brooder height with age, and where several brooders are required they should be hung on adjustable chains from an overhead track so that they can be moved both vertically and horizontally to alter the relative pattern and distribution of heat.

Employees responsible for environment and husbandry should be trained to use appearance, behaviour and noise to control the brooding environment. For example, when chicks appear comfortable during the day but have wet streaks on their back and flanks, it indicates that they have probably been crowding at night, and it is necessary either to increase brooder temperature and alter the distribution of heat or to reduce air speed.

Providing well-insulated accommodation, controlling ventilation (see Chapter 3, this volume) and limiting space per bird all help to reduce heat loss

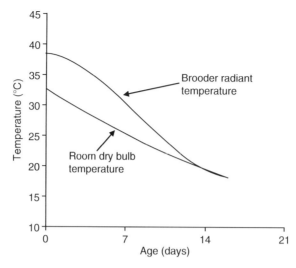

Fig. 4.31. The relationship between age and average brooder and room temperature measured at bird level at 1 and 4 m from radiant-type gas brooder designed to accommodate about 500 ducks providing 2.64 and 1.2 kW/h, respectively, at maximum and minimum heat. Temperatures were recorded in a commercial brooding location holding about 5000 ducks placed at 15 birds per m²: minimum ventilation was provided by electric fans controlled by interval timer to maintain relative humidity below 85% and ammonia at less than 10 ppm.

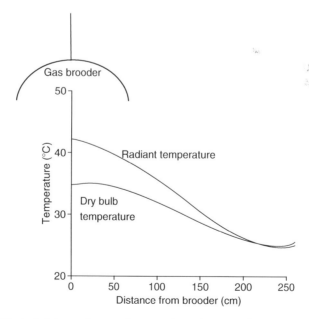

Fig. 4.32. Effect of distance from radiant gas brooder on radiant and dry bulb temperature measured at bird level with brooder located at manufacturers' recommended height from floor and providing 2.64 kW/h of heat. Location and brooder as in Fig. 4.31.

and energy required for heating. Figure 4.31 shows the effect of these factors on room temperature recorded in an intensive brooding facility. However, ducks are hardy and when brooded in relatively small numbers or in simple accommodation can tolerate room temperatures as low as 15°C from day-old, provided there is sufficient brooding heat for them to sleep and rest in a warm environment and feed and water is located close to the brooder. Chicks can be allowed access to shallow bathing water as early as 4 days in summer in temperate climates (see Fig. 3.18) and year-round in the tropics, but it is important to limit water depth to about 4 cm until down has grown sufficiently to provide a waterproof barrier. Birds can be given access to outside pens when ambient temperature in deep shade is above about 20°C, but it is important to provide shade and protection from predation by birds, animals, snakes and occasionally man.

Figure 4.33 provides an estimate of minimum ventilation requirements for table duckling reared intensively on litter and wire, based on unreported large-scale replicated trials in a temperate climate, although requirements can be affected by climate, accommodation and system of husbandry. In cold and temperate climates, recirculating pressure jets (see Chapter 3, this volume) are effective for mixing cold incoming air with warm house air and providing an even distribution of relatively warm fresh air within the accommodation. However, it is important to run recirculation fans at low speed because air speeds at bird level in excess of about 0.2 m/s between day-old and about 7 days will cause crowding and wet down.

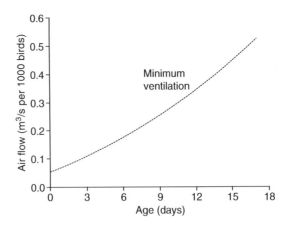

Fig. 4.33. The relationship between age and minimum ventilation requirement per 1000 ducks, measured as inlet volume (m³/s), for birds housed intensively in a temperate climate on wire and litter. Birds were stocked at about 12 birds per m² with air supplied through pressure jets and ventilation provided using 500 mm diameter 1350 rpm diaphragm mounted fans operating at a static pressure of less than 50 N/m². Maximum ventilation (1.2 m³/s per 1000 birds) at 15 days of age was sufficient to maintain house temperature within about 3°C of maximum summer ambient temperature.

To prevent litter becoming saturated and dirty, fresh litter should be spread over the entire floor area each day and minimum ventilation from day-old must be sufficient to prevent relative humidity from exceeding about 85%. This is usually more than sufficient to maintain carbon dioxide at a safe working level of less than 0.3%.

Young chicks are sensitive to ammonia, and empirical evidence suggests that prolonged exposure to levels above about 10 ppm between day-old and 21 days can affect health and performance. It is important to monitor ammonia and adjust minimum ventilation before spreading fresh litter, which can temporarily reduce the level of ammonia by as much as 40%. The installed maximum ventilation should be sufficient to remove heat from litter fermentation and metabolic activity and maintain house temperature at not more than 2–3°C above ambient temperature in hot weather (see Fig. 4.33).

Rearing Environment

To reduce heating costs in cold and cool temperate climates, birds are frequently brooded in about a third of the overall floor area, separated from the remainder of the accommodation by a plastic curtain or insulated and folding partitions. An alternative is to rear birds in custom-designed brooding accommodation, and either herd or transport ducks on custom-designed low-loading trailers to rearing accommodation or range. Feed should be removed 2–3 h before moving, and birds should be herded or transported as early in the day as possible to allow them to settle into their new accommodation before nightfall; it is most important to ensure that birds are actively drinking water before providing feed.

Figure 4.34 describes the minimum ventilation required when rearing birds from about 17 days to slaughter to maintain ammonia below about 15 ppm for birds housed intensively on litter. Figure 4.35 shows that when ambient temperature was about freezing, house temperatures were 7°C and 10°C for birds given minimum ventilation aged 21 and 35 days, respectively, although house temperature will be affected by live weight, minimum ventilation setting, insulation and litter temperature.

There is no economic advantage in reducing house temperature below 12°C (see Fig. 4.3) and litter becomes wet and dirty when house temperature falls below about 8–10°C for more than 2 or 3 h daily, affecting health and feathering, and down quality. In climates where ambient temperature in winter regularly falls below freezing, it may be necessary either to provide supplementary heat or to use a partition to limit floor space until birds are feathered. In cold climates, it may also be advisable to increase the number of birds per square metre to about 7.5–8 birds per m² and spread fresh litter over the floor area twice daily until birds are about 30 days of age, to encourage litter fermentation and increase litter temperature to provide birds with a warm dry bed.

Birds should be provided with a constant supply of drinking water, but drinker design affects water use (see Fig. 4.36 and Chapter 3, this volume) and

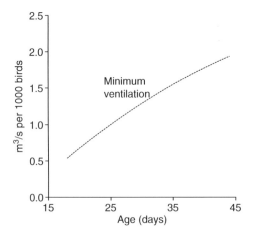

Fig. 4.34. Relationship between age and minimum ventilation requirement per 1000 birds measured as inlet volume (m³/s) for birds housed intensively in a temperate climate on wire and litter. Birds were stocked at 6 birds per m² with air supplied through pressure jets and sidewall inlets and ventilation was provided by 630 mm, 920 rpm diaphragm mounted fans operating at a static pressure of less than 50 N/m². Maximum ventilation (3.6 m³/s per 1000 birds) was sufficient when birds were 43 days and live weight was about 3.3 kg to maintain house temperature within 2–3°C of maximum summer ambient temperature.

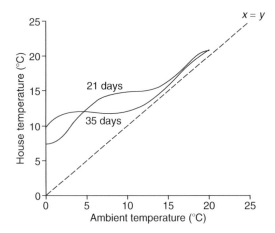

Fig. 4.35. Relationship between ambient and house temperature for birds of the same genotype aged 21 and 35 days with live weights of about 1.25 and 2.52 kg. Birds were placed at 6 birds per m² and given minimum ventilation as shown in Fig. 4.34, with maximum ventilation controlled by thermostat set at 15°C and 12°C, respectively. The roof and walls were insulated to provide a U-value of about 0.6 and 0.75 W/m².°C.

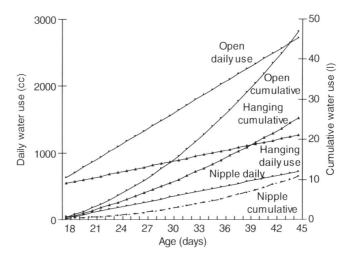

Fig. 4.36. The relationship between age and water use for Pekin with a mature weight of about 3.8 kg given either open troughs, turkey-type hanging plastic drinkers or nipple drinkers.

subsequent volume of effluent. Unreported trials showed that hanging plastic turkey or broiler drinkers reduced feather yield by about 15%, and affected down quality compared to birds given open troughs. Drinkers should be located over suitable drainage (see Chapter 3, this volume) and effluent discharged into approved drainage or irrigated over arable land.

Variability of Performance

A minimal coefficient of variation for live weight within a flock is about 7%, but environmental factors can substantially increase this. Figure 4.37 shows the effect of placing 750 and 500 birds per brooder on live weight distribution at 42 days for birds of the same genotype given similar feed and accommodation. Empirical evidence in different environments shows that within the flock the coefficient of variation is usually about 10–12%, but can be as high as 15%.

Figure 4.38 describes the effect of environment on variation in mean live weight for commercial flocks of the same genotype given similar nutrition, space allowance and temperature and slaughtered over about 2 calendar months at 42 or 43 days of age. Figure 4.39 shows the variation in average live weight for birds slaughtered between 42 and 46 days measured as a coefficient of variation was about 4%; live weight increased linearly by 85 g per day, but weight for age and rate of gain would both be affected by temperature and space allowance (see Figs 4.18 and 4.24). Increasing kill age and nutrient concentration are widely used to overcome the effects of temperature on feed intake and subsequent growth to achieve market weight in summer.

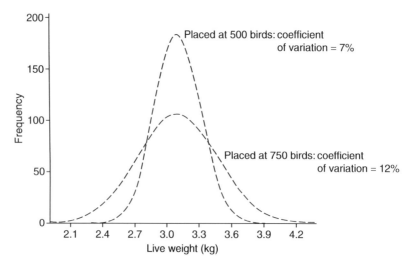

Fig. 4.37. Effect of overcrowding during brooding on weight distribution of 1000 birds of one genotype given similar nutrition and slaughtered at 42 days.

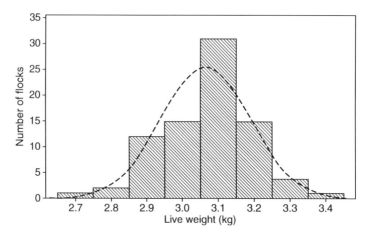

Fig. 4.38. Distribution of mean live weight for flocks of 5000–50,000 slaughtered at 42 or 43 days. Overall mean live weight 3.065 kg, standard deviation 0.125 kg, range 0.68 kg.

Figure 4.40 shows that flock size had no significant effect on mean live weight, although smaller flocks showed much greater variation in performance. Figure 4.41 shows that mortality increased and downgrades decreased with increasing flock size, but overall loss, including both mortality and downgrades, was unaffected by flock size.

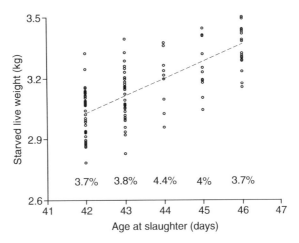

Fig. 4.39. The effect of age at slaughter and environment on mean live weight, and the coefficient of variation, for flocks of 5000–50,000 birds of the same genotype, given feed of the same nutrient specification and reared intensively on litter during winter.

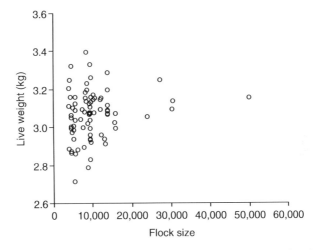

Fig. 4.40. The relationship between flock size and live weight at 42/43 days. Genotype, nutrition and accommodation as in Fig. 4.39.

The History of Improving Quality

Companies in many industries with an international reputation for quality credit their success to the application of concepts proposed by Shewhart (1931), who demonstrated that by constantly monitoring a production process it was possible to reduce variation, improve quality and reduce waste, thus decreasing costs while increasing customer satisfaction. Following the Second World War,

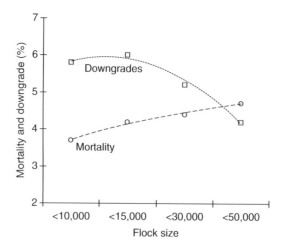

Fig. 4.41. The relationship between flock size, mortality and downgrades for birds slaughtered at 42/43 days. Genotype, accommodation and nutrition as in Fig. 4.39.

General Douglas MacArthur recruited Dr W. Edwards Deming who had worked with Shewhart at the Bell telephone company to assist with the reconstruction of Japanese industry. Deming met with leading Japanese industrialists and suggested that 'statistical process control' offered the potential to assist Japanese industry to reduce costs and improve quality by moving away from monitoring output, to controlling input. He proposed that the causes of variation in performance could be divided into:

1. Common cause – sources of random variation caused by genotype and environment that are always present. When a process is subject to only common cause variation, the process is in a state of statistical control; variation can be measured and predictions made about the expected level of performance. However, this does not mean performance is good or bad, but confirms the process is stable, the causes of variation lie within the system and no amount of adjustment or effort by the workforce can reduce this source of variation. Only management can reduce common cause variation by either altering systems of housing and environment, or by changing feed specifications, vaccination programmes and similar factors.

2. Special cause – variation that occurs intermittently caused by poor husbandry, machinery and equipment failure, inadequate control of heating and ventilation, insufficient bedding, failures in feed and water supply, and similar factors, which increase variation beyond the range that occurs randomly through common cause variation. Special cause variation can usually be identified and eliminated by the stockperson or the hatchery or factory technician.

Deming (1982) suggested that statistical process control offered an opportunity to distinguish between these two types of variation and could assist managers and working staff to identify appropriate action to reduce variation and improve both quality and performance. He described a three-stage programme (see Fig. 4.42):

1. Gain control to produce a stable and predictable performance by eliminating special cause variation.
2. Reduce common cause variation by improving systems, working methods and environment.
3. Constantly seek to reduce waste in all its forms and increase value.

The concepts first proposed by Shewhart and introduced into many industries by Deming are generally credited with substantially improving both quality and economic performance in environments as diverse as aerospace and hospitals and are widely used in many industries to continuously improve both quality and economic performance.

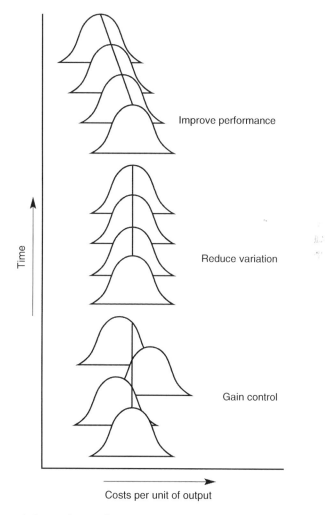

Fig. 4.42. Steps to improving performance.

Traditionally, the poultry industry has used final inspection as a method of achieving quality, but inspecting quality of a product in this manner assumes defects will be produced. In fact, they are expected, but achieving quality in this manner substantially underestimates the resources invested in inferior quality. An alternative approach proposed by Shewhart (1931), Deming (1982, 1986), Juran (1993) and others is to concentrate on prevention: control the input and system of production, rather than achieve quality by post-process inspection.

Process Control

Selection for improved rate of gain and reduced age at slaughter to achieve market weight (see Chapter 9, this volume) have been the principal avenues for improving economic performance of table duckling (see Chapter 2, this volume). However, gradually reducing age at slaughter provides birds with progressively less time to recover from special cause variation. Achieving market weight while reducing age at slaughter depends upon eliminating special cause variation and providing an environment along with sufficient nutrition to enable birds to achieve weight for age.

Special cause variation occurs intermittently, and while it is possible to reduce special cause variation, it is seldom possible to eliminate this source of variation entirely. Preventing special cause variation depends on:

- Continuously training working staff in working methods and routines;
- Providing process control charts to assist working and supervisory staff to identify when special cause variation is affecting performance;
- Training process workers to use charts to recognize when special cause variation is affecting performance, and to identify and rectify the cause.

Control charts for identifying when a special cause is affecting performance are based on the concept that measurement of factors such as live weight produces a pattern of variation (see Figs 4.37 and 4.38) known as the 'normal distribution', and mean values of samples also tend towards this predictable pattern. Using a parameter known as the standard deviation, it is possible to calculate control limits for one or both sides of the process mean, and when sample values fall outside these limits, it provides warning that a special cause may be affecting performance. An alternative explanation is to consider process control in terms of noise (common cause variation) and signal (special cause variation): when overall variation increases beyond what might be expected from common cause variation which is both stable and predictable, it provides a 'signal' that a special cause of variation may be affecting performance; process control charts provide a practical way to distinguish signal from noise.

Charts can be single or double-sided with linear or curvilinear target means (Johnson and Tissel 1990; Coleman *et al.*, 1996; Oakland, 1998). Figure 4.43 shows a double-sided process control chart. Samples are drawn from the population and values entered on to the chart. As long as sample values are located inside the warning lines, the process is in control. However, where values fall outside the limits, it provides warning that a special cause may be affecting the process, and where a second sample confirms this result, it is highly likely a special cause is affecting performance.

Control limits should not be confused with tolerance, or specification limits which express the required or desired quantity or quality. A process can be in control, providing a stable and predictable performance, and yet producing a product outside specification or market requirement.

The presence of unusual patterns, runs, trends or recurring patterns, even when all sample means are within the warning lines, can sometimes indicate a change in process average (Pitt, 1994), and provides early warning of an impending problem, and indicates some action may be necessary before plotted values exceed the warning limit. The following are common patterns (see Fig. 4.44) indicating when special cause is affecting a process:

1. A single point outside either action limit;
2. Eight points forming a run on one side of the centreline;
3. Five or six points forming a trend in one direction;
4. Two points just inside the upper or lower warning limit.

When introducing process control to assist working staff identify the presence of special causes of variation, it is important to obtain the assistance of an experienced statistician to advise on sampling, charting and use of suitable software for recording and charting performance. Charts should be located at the point of production, with farm staff responsible for both sampling and recording sample data onto process control charts, or into a computer terminal, and investigating when performance exceeds control limits.

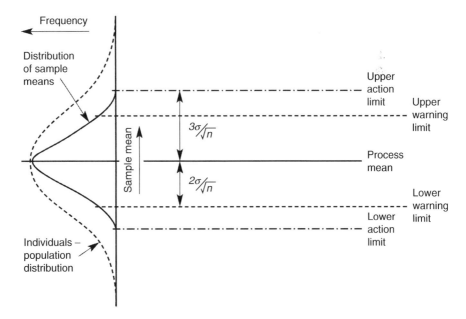

Fig. 4.43. Principle of mean control chart. (Reprinted from Oakland, 1998. With permission of Butterworth-Heinemann.)

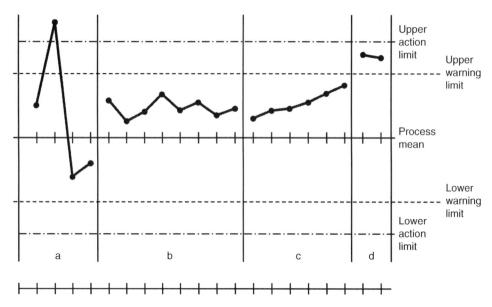

Fig. 4.44. Process control chart. (Modified from Pitt*, 1994.)

A convenient way to make charts readily understandable is to use the concept of traffic lights to illustrate the principle of process control, providing charts coloured green, amber and red, indicating 'process is in control', 'proceed with caution' and 'investigate the process', respectively. When a process is in control, performance is stable and predictable; data recorded on the chart should be scattered in the green zone on either side of the process target mean, although in large organizations when monitoring the performance of hundreds of farms, a sample mean will occur by chance in the amber zone, outside the warning limits, about three or four times a week, and beyond the action limits in the red zone, about once a week. Where a chart indicates a special cause is affecting performance, it will occasionally be sufficient to adjust the process, but real improvement in performance can only be achieved when special cause variation has been removed.

Reducing Common Cause Variation and Improving Performance

No agricultural process can produce consecutive items which are identical in weight, composition or appearance because of the effects of factors such as genotype, system of production, environment, furniture and equipment, nutrition and health programmes. Variation of this nature is known as 'common cause' or 'system variation'. This type of variation is always present and can only be reduced by changing or altering systems, environment, and nutrition (see Chapter 5, this volume) and health programmes.

Environmental variables affecting performance can be divided into those such as daylength and light intensity, which can be modified to improve performance, and those such as high environmental temperature, which affect performance in a predictable manner, but cannot for practical or economic reasons be controlled.

The first step towards improving performance is to measure the opportunity. For example, when considering weight for age, it is important to compare current commercial performance with growth and feed intake for the same genotype given non-limiting nutritional and environmental conditions.

The next step is to quantify the separate effects of nutrition (see Chapter 5, this volume) and environment on performance. The economic effects of nutrition can be measured using marginal cost–benefit analysis (Dillon, 1977) to review the marginal cost of increasing nutrient concentration in relation to the marginal benefit of improved performance. The effects of environment on physical and economic performance can be measured both within and between locations. Recording feed intake, live weight, feather quantity and quality, ventilation and gas concentrations, utility costs, and processed quality and yield provides an opportunity to measure the economic effects of both discrete and compound environmental variables on physical and economic performance.

A portable industrial hoist along with an accurate strain gauge load cell was used for lifting and weighing free-standing feeders and measuring feed intake in the majority of trials described in this and subsequent chapters; live weight was recorded using a custom-designed, top-loading and side-emptying crate, suspended from an accurate strain gauge load cell.

Measuring feed intake and live weight in this manner provided accurate records of diverse environmental treatments in different locations and climates. Subsequent analysis of these data provided descriptions of the effect of environmental treatments on relative feed and nutrient intake and growth pathways; this, along with information on feather, processing quality, utility costs and similar factors provided the opportunity to substantially reduce costs and improve both physical and economic performance.

References

Cherry, P. (1993) Sexual maturity in the domestic duck. PhD thesis, The University of Reading, Reading, UK.

Coleman, S., Greenfield, T., Jones, R., Morris, C. and Puzey, I. (1996) *The Pocket Statistician A Practical Guide to Quality Improvement*. John, London.

Deming, W.E. (1982) *Quality, Productivity and Competitive Position*. Centre for Advanced Engineering Study, MIT, Cambridge, Massachusetts.

Deming, W.E. (1986) *Out of the Crisis*. Centre for Advanced Engineering Study, MIT, Cambridge, Massachusetts.

Dillon, J.L. (1977) *The Analysis of Response in Crop and Livestock Production*. Pergamon Press, Oxford.

Emmans, G.C. and Charles, D.R. (1977) Climatic environment and poultry feeding in practice. In: Haresign, W., Swan, H. and Lewis, D. (eds) *Nutrition and the Climatic Environment*. Butterworths, London, pp. 31–48.

Gonzalez, D.A. and Marta, B. (1980) Duck breeding in Venezuela. *Tropical Animal Production* 5(2), 191–194.

Johnson, L.T. and Tissel, J. (1990) *A More Capable Production*. Chartwell-Bratt, Bromley, UK.

Juran, J.M. (1993) *Quality Planning and Analysis*, 3rd edn. McGraw-Hill, New York.

Maruyama, K., Vineyard, B., Akbar, M.K., Shafer, D.J. and Turk, C.M. (2001) Growth curve analyses in selected duck lines. *British Poultry Science*, 42, 574–582.

Mount, L.E. (1979) *Adaptation to Thermal Environment*. Edward Arnold, London.

Nielson, S. (1992) *Mallards*. Swan Hill Press, Shrewsbury, UK.

Oakland, J.S. (1998) *Statistical Process Control A really Practical Guide* 3rd edn. Butterworth-Heinmann, Oxford.

Pitt, H. (1994) *SPC for the Rest of Us – A Personal Path to Statistical Process Control*. Addison–Wesley, Reading, Massachusetts.

Queeny, E.M. (1983) *Prairie Wings, The Classic Illustrated Study of American Wildfowl in Flight*. Dover Publications, New York.

Shewhart, M.A. (1931) *Economic Control of Quality of Manufactured Product*. Van Nostrand, New York.

Testik, T. (1988) A study on some characteristics of the Pekin Duck. In: *Proceedings of the International Symposium on Waterfowl Production*. The Satellite Conference for the 18th World's Poultry Congress, Beijing, China. Peragamon Press, Oxford, pp. 86–89.

5 Nutrition and Factors Affecting Body Composition

Dietary Energy

When ducklings are offered diets with different energy concentrations, they respond by eating less of the higher-energy feed. However, this reduction in intake does not fully compensate for the higher-energy concentration and ducklings gain weight faster on high-energy feeds. The responses, which are similar to those reported for chicks by Fisher and Wilson (1974), are illustrated by three unreported trials shown in Figs 5.1 and 5.2. Increasing dietary energy reduced cumulative feed intake by about 230 g, but increased live weight by 70 g, for each megajoule per kilogram increase in dietary metabolizable energy (ME). Figure 5.3 shows that despite differences in genotype and climate all trials showed a similar marginal gain in live weight of about 16 g/MJ increase in cumulative ME intake to 42 days.

Increasing dietary energy improved the efficiency of feed conversion in all trials, and analysis of the results in Fig. 5.4 provides the following estimate for the effect of dietary energy (x, MJ ME/kg) on efficiency of feed conversion to 42 days of age.

Feed conversion ratio (FCR) = $6.203 - 0.549x + 0.018x^2$

Increasing dietary energy also increased yield (eviscerated weight measured as a percentage of plucked body weight) in two trials (see Fig. 5.5) designed to investigate the effect of energy to protein ratio on body composition and reported later in this chapter. However, this increased yield is only a reflection of the effect of dietary energy on live weight. Analysis of the effect of live weight on eviscerated weight shows that yield, measured at 47 days, improves by about 0.7% for each 100 g increase in plucked body weight. Unreported serial slaughter trials in both the UK and the USA (M.S. Lilburn, Ohio State University, 2006, personal communication) confirm that increasing live weight improves yield linearly by about 0.25% per day between 38 and 50 days.

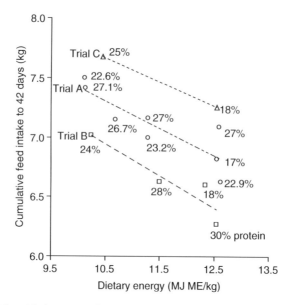

Fig. 5.1. Relationship between dietary energy concentration and feed intake to 42 days recorded in three trials. Protein contents of the diets varied (as indicated) but this did not have any consistent effect on feed intake.

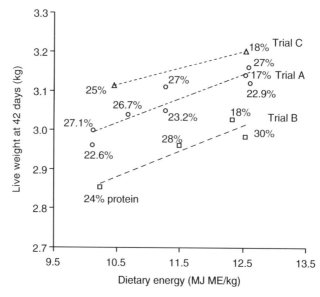

Fig. 5.2. Relationship between dietary energy concentration and growth to 42 days recorded in the three trials reported in Fig. 5.1.

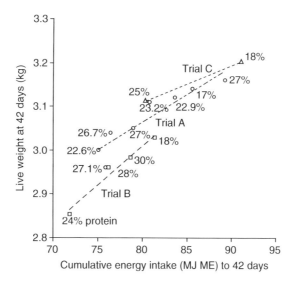

Fig. 5.3. Relationship between cumulative energy intake and live weight at 42 days recorded in the three trials described in Fig. 5.1.

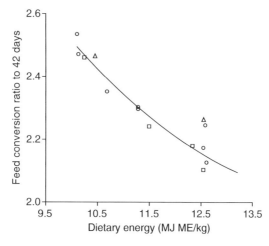

Fig. 5.4. Relationship between dietary energy and feed conversion ratio (FCR) recorded in three trials described in Fig. 5.1.

A trial designed to investigate opportunities to improve both physical and economic performance by increasing dietary energy and reducing protein after 35 days (see Fig. 5.6) showed that increasing dietary energy to 14 MJ/kg reduced subsequent feed intake by about 10%, or 25 g per bird day. However, birds given either a 12.6 or a 14 MJ/kg diet achieved similar intakes of energy to 48 days. Reducing protein to 11% at 35 days for birds given an increase in

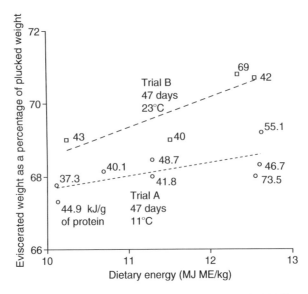

Fig. 5.5. Effect of dietary energy and energy to protein ratio (kJ ME/g protein) on eviscerated weight measured as a percentage of plucked weight recorded in two trials for birds of similar genotype. Birds were reared intensively on litter and average trial temperatures for trials A and B were 11°C and 22°C, respectively. The energy to protein ratio of diets varied (as indicated) but this did not have any consistent effect on eviscerated weight measured as a percentage of plucked body weight.

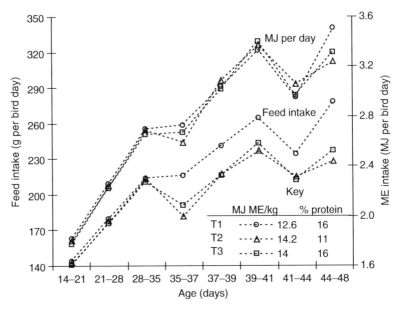

Fig. 5.6. The effect of increasing dietary energy and reducing protein from 35 days on subsequent feed intake. Birds were reared on a 12.6 MJ ME/kg and 18% protein diet to 35 days and then given three dietary treatments as shown in the table.

energy had no effect on feed intake, but substantially reduced rate of gain from 35 to 48 days of age (see Fig. 5.7).

Another trial with the same genotype investigated opportunities to increase energy intake by including molasses in the water supply from 35 days of age. Figure 5.8 shows that feed intake immediately declined by about 10%, but birds preferred drinking fresh water and consumption of the sugar solution gradually

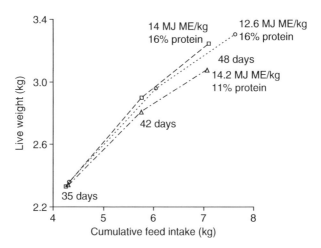

Fig. 5.7. Effect of increasing dietary energy and reducing protein at 35 days on cumulative feed intake and growth to 48 days; the same trial as in Fig. 5.6.

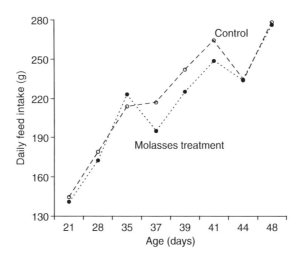

Fig. 5.8. Effect of feeding molasses from 35–48 days on daily feed intake. At 35 days birds in three pens were given a 5% solution of molasses for 10 h per day and were provided with clean fresh water for the remaining 14 h. The control pens received water at all times.

declined and feed intake increased. Providing energy in the water supply in this manner reduced feed intake and final live weight by about 5% and 3.8%, respectively. Further trials providing other soluble sugars such as maize syrup in the water supply, aimed at increasing energy intake when average ambient temperature exceeded 25°C, also reduced feed intake without improving physical performance.

Controlled Energy Intake

In China, forced feeding has been traditionally used to provide a relatively fat carcass for the 'Pekin roast duck' restaurant market. Figure 5.9 shows the typical growth and feed intake for birds fed *ad libitum* to 6 weeks and subsequently force-fed to 8 weeks with a moderately high-energy but low-protein diet (12.6 MJ/kg and 12% protein diet). This substantially increases fatness. However, to satisfy the ever-growing demand for 'Pekin roast duck' and improve economic performance, birds are increasingly reared intensively on wire and litter without access to swimming water. They are no longer force-fed, but fed *ad libitum* on pelleted feed of increased nutrient concentration. This gives a similar live weight at 8 weeks to force-fed birds.

Increasing consumer preference for lean meat in the European Community (EC) and USA has encouraged some duck producers to consider using feed restriction and an increased age at slaughter to reduce body fat and increase breast and leg meat. However, restricting feed intake necessarily reduces rate

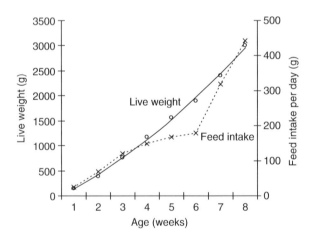

Fig. 5.9. Relationship between age, feed intake and growth for birds given three meals in the daytime, and *ad libitum* feed at night of a 16% protein concentrate and green feed diet (concentrate to green feed ratio 1:4.5). From 6 weeks of age, birds were force-fed four times daily a 12.6 MJ ME/kg and 12–13% protein concentrate mixed with water. Birds were given 150 g per day of concentrate, increasing daily by about 25 g to provide about 325 g by 7 weeks of age. (Data from Jung and Zhou, 1980.)

of gain. In a trial where birds were restricted to 75% or 85% of the feed for birds fed *ad libitum* (see Fig. 5.10), ducks required 68 and 56 days, respectively, to reach the live weight that the birds fed *ad libitum* achieved at 47 days. This adversely affects efficiency of feed conversion and cost of production. The effects of age at slaughter, feed restriction and other factors on body composition are described later in this chapter.

Factors Affecting Response to Dietary Energy

As with all homeotherms, the energy needed by a duck to maintain a constant body temperature is related to its surface area, but reduces as environmental temperature increases. Figure 5.11 describes the effect of body weight on maintenance energy requirement of mature males maintained at 10°C and 26°C. A regression analysis of ME intake on live weight (*W*) provides estimates of maintenance requirement of 583 kJ/kg $W^{0.75}$ day at 10°C and 523 kJ/kg $W^{0.75}$ day at 26°C for genotypes selected for efficiency of feed conversion and body composition.

Modern commercial hybrids in the USA and EC are usually derived from male lines selected for growth and efficiency of feed conversion mated to female lines selected for breeding performance, body composition and similar traits. Figure 5.12 illustrates the effect of genetic selection for efficiency of feed

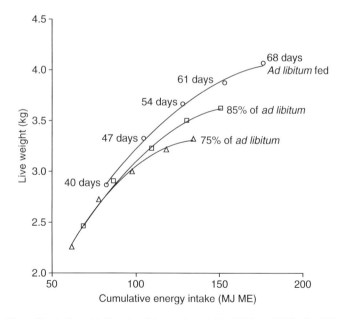

Fig. 5.10. The effect of restricting feed from day-old to 75% or 85% of *ad libitum* intake on cumulative energy intake and live weight. Birds were given a diet with 12.7 MJ ME/kg and 22% protein. (Data from Cherry, 1993.)

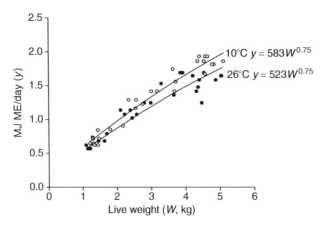

Fig. 5.11. Estimates of maintenance requirement for drakes held at 26°C and 10°C. Mature drakes of seven Pekin genotypes were selected to give a wide range of live weights. Birds were housed on litter in heated accommodation in individual pens and given a fixed feed allowance of about 90% of their *ad libitum* feed intake. Temperature in the house was thermostatically controlled at 26°C for the first 16 weeks and then reduced to 10°C for a further 8 weeks. The birds were weighed at regular intervals and the live weight at the end of each period was adopted as the weight maintained by the feed allocated. (Data from Cherry and Morris, 2005.)

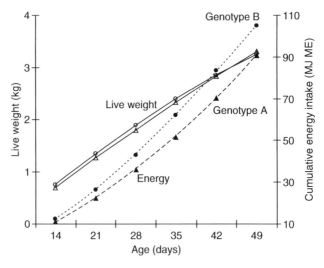

Fig. 5.12. The relationship between energy intake and growth for two Pekin genotypes of similar mature live weight. Genotype A was selected for body composition and efficiency of feed conversion, and B was an unselected genotype. Birds were reared intensively on litter with an average trial temperature of 13°C and given a starter diet to 2 weeks and then a 12.8 MJ ME/kg and 18.5% protein diet.

conversion and body composition upon the energy requirement to 49 days in genotypes of similar mature body weight. Figure 5.13 shows a comparison of a genotype selected for both rate of gain and efficiency of feed conversion, with one selected for breeding performance. In this comparison genotype B has a lower growth rate and lower mature body size. Selection for growth and efficiency has reduced age at slaughter by about 9 days and cumulative energy requirement to market weight by about 30%. However, the poor reproductive performance of genotype A means it cannot be used as a pure line for commercial production. By mating drakes of line A to ducks of line B a commercial operation can have the egg production and hatchability of line B combined with growth and efficiency in the progeny which is halfway between the two lines.

Figure 5.14 describes the effect of temperature on energy intake and subsequent growth. Analysis of the data provides the following estimate for the effect of cumulative energy intake (x, MJ ME) to between 46 and 48 days on growth.

$$\text{Live weight (kg)} = -3.624 + 0.1242x - 0.00054x^2$$

In a trial attempting to increase energy intake and subsequent growth at high temperature and humidity by increasing dietary energy (Fig. 5.15), it was found that birds reduced their feed intake when given the higher-energy diet from 28 days, but achieved about the same daily energy intake and live weight at slaughter as birds given the normal diet. A similar response was described for birds

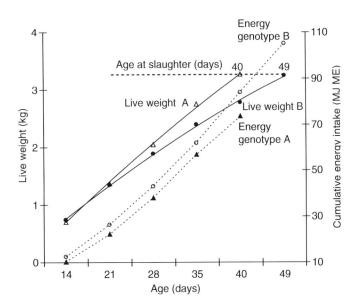

Fig. 5.13. Energy consumption to similar weight at slaughter for two Pekin genotypes with different mature live weights selected for (A) growth and efficiency of feed conversion and (B) breeding performance.

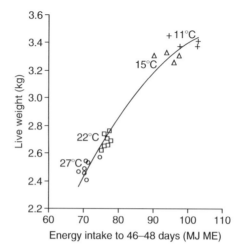

Fig. 5.14. Relationship between prevailing environmental temperature, energy intake and growth to 46–48 days for Pekin of the same genotype reared on a starter diet to 18 days and then given diets containing 12.4–13.2 MJ ME/kg to 46–48 days. Birds in four separate trials were reared on litter in the same open-sided and naturally ventilated accommodation. Minimum and maximum temperatures were recorded daily in each trial to obtain the trial average temperatures of 11°C, 15°C, 22°C and 27°C.

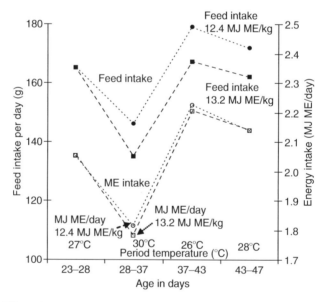

Fig. 5.15. Effect of age, high ambient temperature and dietary energy on daily feed and energy intake from 28 to 48 days of age. Birds were reared on litter in open-sided accommodation in a hot and humid climate and given a diet containing 12.4 MJ ME/kg and 18% protein from 14 to 28 days, and then given diets with 18% protein and either 12.4 or 13.2 MJ ME/kg. Maximum and minimum temperatures were recorded daily to provide an estimate of average temperature for each period.

reared in a temperate climate (Fig. 5.6), confirming that birds regulate their feed intake primarily to control intake of energy. In the trials reported in Fig. 5.1, diets ranging from 10 to 12.5 MJ ME/kg were compared (in a temperate climate) and, over that range, birds given the lower-energy diets did not succeed in raising their feed intakes sufficiently to maintain energy intakes to match those of birds on 12.5 MJ ME/kg. The evidence presented in Chapter 4 (this volume) along with data recorded in reported trials supports the conclusion that increasing 'effective' temperature reduces the opportunity for birds to lose heat; they reduce metabolic activity and regulate intake of energy to prevent hyperthermia and maintain homeostasis, and this limits their growth.

Protein, Amino Acids and Performance

Figure 5.16 describes the effect of increasing dietary protein in seven trials on growth to between 42 and 48 days. Increasing protein over the range 14–24% increased live weight in a curvilinear manner, with genotype, temperature and dietary energy probably responsible for differences in response between trials. Figure 5.17 shows increasing protein in six trials lowered the FCR in a roughly linear manner by about 1.3%, for each 1% increase in protein.

In countries where protein is scarce or expensive, ducks are frequently reared on low-protein feeds. Figure 5.18 shows that feeding a 15.7% protein diet from day-old to 42 days substantially reduced feed intake to about 40 days and rate of gain to 28 days. Figure 5.19 shows that feeding the low-protein starter diet delayed growth to both 14 and 42 days by about 3 days. Evidence

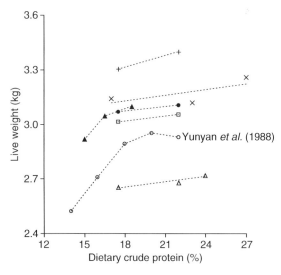

Fig. 5.16. The effect of increasing protein in diets fed from 14 days to slaughter at 42–48 days on growth recorded in seven trials for Pekin of different genotypes reared in different climates. (Data from unreported trials and Yunyan *et al.*, 1988.)

Fig. 5.17. The effect of increasing dietary protein on efficiency of feed conversion recorded in six trials for different Pekin genotypes reared in temperate and hot climates and slaughtered at between 42 and 48 days.

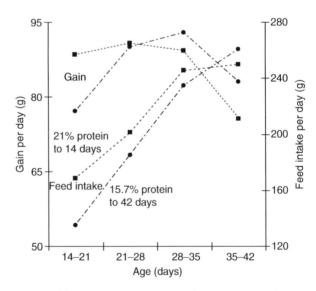

Fig. 5.18. The effect of feeding a low-protein diet from day-old on feed intake and rate of gain. Birds were given either 12.9 MJ ME/kg and 15.7% protein from day-old to 46 days or 12.9 MJ ME/kg and 21% protein to 14 days followed by 12.9 MJ ME/kg and 15.7% protein to 42 days of age. Birds were reared in open-sided accommodation in a hot climate.

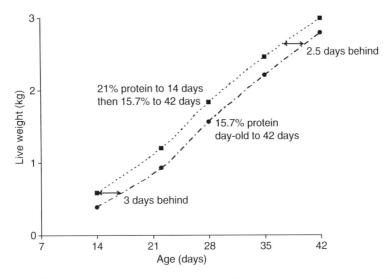

Fig. 5.19. Effect of dietary protein on growth to 42 days. Diets and accommodation as in Fig. 5.18. FCR to 42 days was 2.244 for birds given 21% protein and 2.255 for 15.7% protein. However, to achieve an equal live weight it would be necessary to increase age at slaughter for birds fed low-protein diets by 2–3 days, increasing their FCR to about 2.31.

will be presented later in the chapter to show that feeding birds a low-protein starter feed can also substantially affect body composition.

Commercial producers occasionally use finisher feed remaining from a previous flock as starter feed. Figure 5.20 shows that feeding birds a low-protein diet (16% protein) to 6, 8 or 10 days reduced early rate of gain. When the birds initially fed low protein were given 21% protein, they grew at a similar rate to birds given 21% protein from day-old, starting from the age and weight achieved when protein restriction was removed. Feeding birds in this manner reduced 48-day live weight (Fig. 5.20) but improved FCR. However, to reach the same weight as birds given 21% protein from day-old it would be necessary to increase age at slaughter by about 1, 2 and 4.5 days for birds given low-protein feed for 6, 8 and 10 days, thus increasing FCR by about 2%, 2.5% and 4.5%, respectively, above the control birds.

Amino Acids

Proteins are a combination of amino acids, and protein quality can be measured by the relative concentration and availability of amino acids provided by a feed ingredient or compound feed. Some amino acids cannot be synthesized by the bird and are known as 'essential' amino acids. Performance depends upon providing birds with a sufficient quantity of essential amino acids, along with an adequate supply of non-essential amino acids. Lysine and the sulphur-bearing

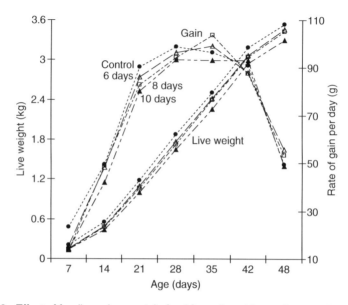

Fig. 5.20. Effect of feeding a low-protein feed from day-old on subsequent growth. Birds were given a 12.6 MJ ME/kg and 16% protein diet to 6, 8 or 10 days and then duck starter (12.9 MJ ME/kg, 21.5% protein), grower (12.6 MJ ME/kg, 17.2% protein) and finisher (12.6 MJ ME/kg, 16% protein) to 17, 35 and 48 days, respectively. Birds reared on the standard feed programme achieved a live weight of 3.535 kg and an FCR of 2.335 at 48 days, and birds given low protein to 6, 8 and 10 days achieved live weights of 3.464, 3.426 and 3.294 kg and FCR of 2.327, 2.31 and 2.277, respectively.

amino acids methionine and cystine are most likely to be undersupplied, and so are described as 'first-limiting' amino acids.

Figure 5.21 describes the results of an unreported trial investigating the effect of available lysine on performance, and shows that increasing the intake of available lysine in a temperate climate improved rate of gain per day for a dual-purpose Pekin genotype between 14 and 35 days, but had no substantial effect thereafter. Analysis of these results provides an estimate of the effect of lysine intake between 14 and 35 days (x, g lysine) on daily rate of gain.

Daily rate of gain (14–35 days, g) = 54.77 + 12.249x

Figure 5.22 shows that increasing dietary concentration of available lysine beyond 0.9% only marginally improved average rate of gain to 42 days of age, and results of a further trial (Fig. 5.23) show that reducing protein and available lysine to 15.5% and 0.7%, respectively, at 30 days had no substantial effect on subsequent rate of gain or efficiency of feed conversion to 42 days, but this response would be affected by genotype, dietary energy and temperature.

Methionine and cystine are also essential amino acids required for growth of muscle and feather. Figure 5.24 describes the effect of increasing the concentration of available methionine and cystine on feather growth in two unreported trials. Increasing the available methionine and cystine from 0.6% to 0.8% increased primary feather growth to 42 days in both trials by about 10 mm or

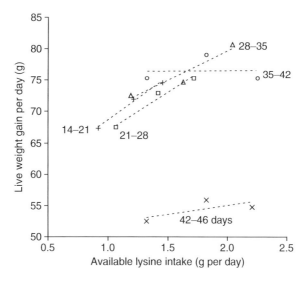

Fig. 5.21. Effects of available lysine intake on growth of Pekin ducks from 14 to 46 days. All birds were given duck starter to 14 days and then provided with diets containing 12.6 MJ ME/kg, 0.9% methionine plus cystine and either 1.1%, 0.9% or 0.7% available lysine. Trial average of daily maximum and minimum temperatures was about 16°C.

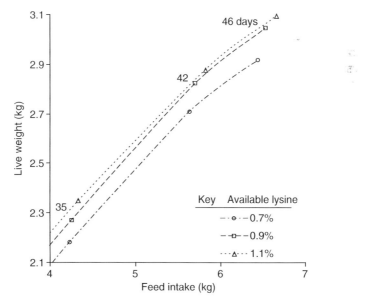

Fig. 5.22. Effect of age and available lysine concentration (%) on feed intake and growth to 46 days. Genotype, diet, accommodation and temperature as in Fig. 5.21.

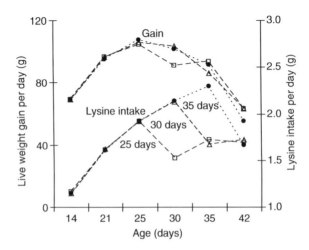

Fig. 5.23. Effect of reducing percentage of dietary protein from 18% to 15% and available lysine from 0.9% to 0.7% at 25, 30 and 35 days on daily lysine intake and live weight gain per day to 42 days for Pekin reared intensively on litter in a temperate climate.

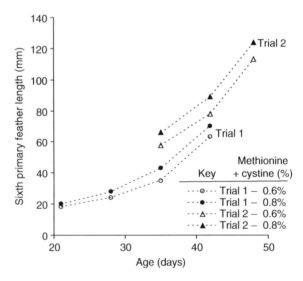

Fig. 5.24. Effect of age and dietary concentration of methionine plus cystine on growth of sixth primary feather recorded with different Pekin genotypes in two trials. Birds were reared intensively on litter in a temperate climate and given duck starter to 14 days, and then diets containing 12.8 MJ ME/kg, 18% protein, 0.85% available lysine and either 0.6% or 0.8% of methionine plus cystine to 42 and 49 days of age, respectively.

14%, and substantially improved general body feathering. It also increased body weight by about 4%.

Energy and Protein Interaction

Increasing dietary energy (see Fig. 5.2) and protein (see Fig. 5.16) can improve growth and efficiency of feed conversion. This raises the question whether the response to protein depends on the level of energy selected, i.e. whether the two inputs interact. Experimental data (Yunyan *et al.*, 1988) show that increasing both energy and protein in the range 11–12 MJ ME/kg and 14–22% protein improved rate of gain to 56 days for a relatively slow-growing Pekin genotype (Fig. 5.25). Analysis of the data provides the following estimate for the effect of dietary energy (*x*, MJ ME/kg) and percentage of protein (*y*) on live weight at 56 days in the trial.

$$\text{Live weight (kg)} = -52.7047 + 8.75925x + 0.342286y$$
$$- 0.36875x^2 - 0.008035y^2$$

There is no evidence of interaction here. Analysis of results from another trial where Pekin selected for growth and efficiency of feed conversion were given diets of either 12 or 13 MJ ME/kg, and 15%, 16.5% or 18% protein also shows (see Fig. 5.26) that increasing energy improved live weight and efficiency of feed conversion by about 2% and 5%, respectively. Increasing protein from 15% to 18% improved growth and efficiency of feed conversion linearly by about 2% and 1%, respectively, per 1% increase in dietary protein (although the response would be curvilinear over a wider protein range). Analysis of the data provides the following estimate for the effect of age (*x*, days), percentage of protein (*y*) and energy (*z*, MJ ME/kg) on live weight and FCR.

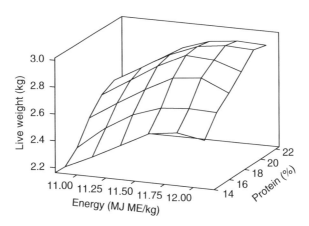

Fig. 5.25. The effect of increasing energy and protein on live weight. (Data from Yunyan *et al.*, 1988.)

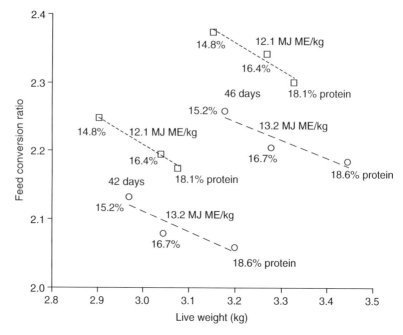

Fig. 5.26. Effect of increasing energy and protein upon growth and efficiency of feed conversion recorded at 42 and 46 days of age. Birds were reared intensively on litter in a temperate climate, and given either 12.1 or 13.2 MJ ME/kg, and about 15%, 16.5% and 18% protein diets from 16 days. Average temperature during the trial was about 23°C.

$$\text{Live weight (kg)} = -1.5261 + 0.0753x + 0.0556y + 0.0347z$$

$$\text{FCR} = 2.570 + 0.0263x - 0.0216y - 0.0911z$$

Again there is no evidence of interaction.

Optimum Nutrient Concentration

Marginal cost–benefit analysis provides a method of identifying optimum economic nutrient concentration as the point where marginal cost of increasing nutrient concentration (for a particular genotype in a given environment) equals the value of marginal return. However, the effect of nutrient concentration on age at slaughter and subsequent processing yield should also be included in the cost–benefit analysis, because Fig. 5.26 shows that providing 18.6% protein and 13.2 MJ ME/kg diet reduced the age at slaughter needed to achieve a market weight of 3.2 kg by about 4 days compared with other diets. This increases productivity per square metre by about 8% per annum and so reduces the fixed costs per kilogram produced, but also affects body composition and processing yield (see 'Age, sex, maturity and body composition and Age at slaughter and body composition,' towards the end of this chapter). All these factors must be taken into account to reach an optimal feeding strategy.

Factors Affecting Response to Nutrient Concentration

High temperatures reduce feed intake and subsequent growth (see Figs 4.3 and 4.4). Many producers in both temperate and continental climates increase nutrient concentration during summer in an attempt to maintain nutrient intake and achieve market weight. However, two unreported trials carried out at high temperature and humidity (see Figs 5.27 and 5.28) show that increasing dietary protein concentration and consequent protein intake improved growth and efficiency of feed conversion, but could not restore normal growth rates because birds are forced to reduce intake of energy to limit metabolic heat output (see Fig. 5.11) and maintain homeostasis. Analysis of the data provides the following estimates for the effect of total protein intake (x, g) on growth to 46 days at 22°C and 27°C, and the effect of dietary protein percentage (y) on efficiency of feed conversion.

Live weight (kg) at 22°C $= -1.983 + 0.0074x - 0.00000291x^2$

Live weight (kg) at 27°C $= -1.082 + 0.0062x - 0.00000266x^2$

Efficiency of feed conversion at 22°C and 27°C $= 6.279 - 0.399y$
$+ 0.0097y^2$

Genotype can also affect the response to nutrient concentration. Figure 5.29 describes the effect of increasing nutrient concentration on feed intake and growth for unselected and selected Pekin genotypes. Increasing nutrient concentration improved efficiency of feed conversion for the selected genotype by

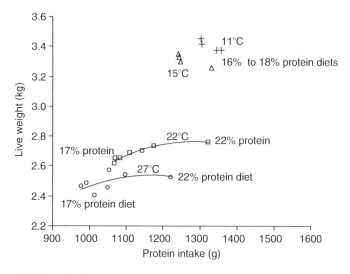

Fig. 5.27. Relationship between prevailing environmental temperature, protein intake and growth to 46 or 48 days for Pekin given starter feed to 18 days and then a range of diets with different protein contents. Birds in four consecutive trials were reared on litter in the same open-sided and naturally ventilated accommodation. Maximum and minimum temperatures were recorded daily to obtain overall trial average temperatures of 11°C, 15°C, 22°C and 27°C.

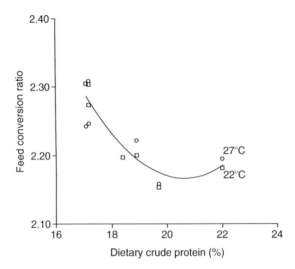

Fig. 5.28. Effect of dietary crude protein on efficiency of feed conversion to 46 days recorded in two trials for Pekin of the same genotype experiencing mean environmental temperatures of 22°C and 27°C, respectively. Genotype, accommodation and diets as described in Fig. 5.27.

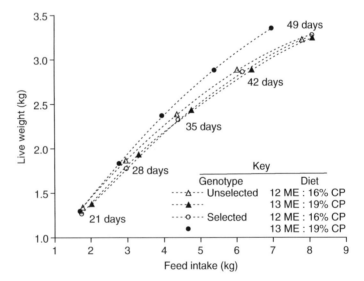

Fig. 5.29. The effect of increasing nutrient concentration from 16 days to slaughter on growth as a function of feed intake for unselected and selected genotypes of similar mature body weight. Birds were reared intensively on litter and given the same starter ration to 16 days, and then diets containing either 12 MJ ME/kg and 16% protein or 13 MJ ME/kg and 19% protein to slaughter at 49 days. Average of daily maximum and minimum temperatures was 15°C.

about 15%, but decreased efficiency for the unselected genotype by about 3%. The selected genotype increased the intake of the lower nutrient concentration diet to satisfy the requirement for protein, adversely affecting FCR, and then stored the surplus energy as fat.

Figure 5.29 provides clear evidence of a genotype–nutrition interaction. Genetic selection for improved efficiency of feed conversion has reduced relative feed and nutrient intake. Achieving genetic potential depends upon providing an adequate supply of energy and nutrients. Providing diets of lower nutrient concentration suitable for indigenous birds will adversely affect productivity of imported selected genotypes in developing countries. Conversely, there is no advantage in increasing nutrient concentration for local genotypes.

Free-choice Feeding

Providing feeds of different nutrient composition gives birds some opportunity to select nutrients on the basis of their metabolic requirement and, by recording feed intake separately at regular intervals, it is possible to measure the effect of age, genotype and environment on nutrient intake (see Figs 5.30 and 5.31). At both temperatures ducks changed their preferences with age, eating more of the

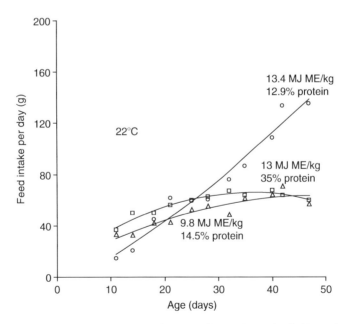

Fig. 5.30. The effect of age on intake of three different feeds for Pekin maintained at 22°C and offered diets providing different concentrations of energy and protein. Birds were reared intensively on litter and the three feeds were supplied in separate feeders from 7 days of age. Consumption was recorded at regular intervals. Indirect gas heating was used to maintain environmental temperature at about 22°C to slaughter.

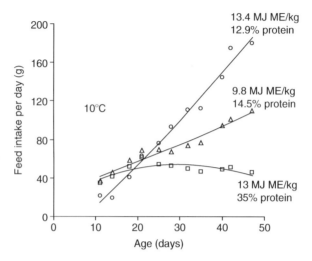

Fig. 5.31. The effect of age on intake of three different feeds for Pekin maintained at 10°C and given diets providing different concentrations of energy and protein. Accommodation and nutrition as in Fig. 5.30.

high-energy feed as they grew older and fatter. Consumption of the 9.8 MJ/kg diet, which contained more indigestible fibre, was notably greater for older ducks at the lower temperature. Intake of the very high-protein diet increased between 10 and 30 days (although forming a steadily smaller fraction of the total feed intake) and then declined after 30 days. Figure 5.32 describes the resulting protein intake expressed as a percentage of total feed intake, and energy intake expressed as kilojoules per gram of protein. At 22°C birds selected an average protein and dietary energy of 20.8% protein and 12.18 MJ ME/kg and, at 10°C, 19.2% protein and 12.03 MJ ME/kg. Analysis of the data provides the following estimates for the effect of age (x, days) and temperature on 'free-choice' dietary percentage of protein and intake (kJ/g) of protein.

At 22°C, protein (%) = $23.096 - 0.03x - 0.0017x^2$

At 10°C, protein (%) = $24.882 - 0.25x + 0.0014x^2$

At 22°C, protein (kJ/g) = $50.713 + 0.107x + 0.0061x^2$

At 10°C, protein (kJ/g) = $44.213 + 0.767x - 0.0023x^2$

Birds reared at 22°C consumed less energy and selected a narrower energy to protein ratio compared to birds reared at the lower temperature. Energy intake is governed by capacity to dissipate heat, but protein intake for birds given a free choice appears to be controlled by their requirement for maintenance and growth, relative to consumption of energy.

Over about the last 25 years selection for growth, efficiency of feed conversion and increased lean mass has increased daily average rate of gain by about 25%, reduced age at slaughter by about 8 days and average daily feed intake by about 12%, but published estimates of nutrient 'requirement' have

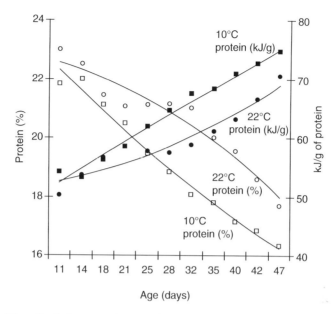

Fig. 5.32. The effect of age on the protein content of the mixture selected from diets offered free-choice and on energy intake per gram of protein. Birds were maintained at either 10°C or 22°C. Feed intake and accommodation as in Figs 5.30 and 5.31.

changed little in that time. However, when unselected and selected geno-types of similar mature body weight were given free-choice diets, similar in composition to those described in Figs 5.30 and 5.31, the unselected genotype consumed an overall average diet composition of 17.8% protein and 12.7 MJ ME kg, whereas the selected birds chose an average diet containing 21% pro-tein and 13.1 MJ ME kg.

Pellets, Mash and Grain Plus Concentrate

The way that food is provided is dictated by cost and custom; the following methods are widely used:

- Compound mash or pellet;
- Cereal grain and pellet concentrate mixture;
- Controlled quantity of concentrate supplied as mash or pellets and either restricted or *ad libitum* cereal grain.

Pellet feeding is popular because, without access to bathing water (see Chapter 3, this volume), mash-fed birds have difficulty in keeping their bills and feathers clean. Figure 5.33 describes the effect of feeding pellets and mash on perform-ance. Feeding mash from 17 days substantially reduced feed intake and rate of gain, but improved efficiency of feed conversion to 48 days of age. However,

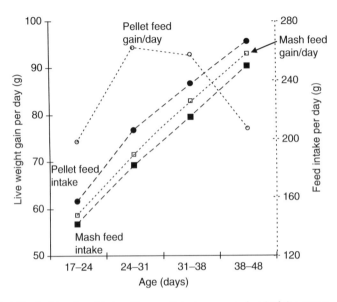

Fig. 5.33. Effect of feeding birds either pelleted or mash feed of the same specification on daily feed intake and rate of gain. Birds were reared intensively on litter and given 3 mm diameter starter pellets to 17 days and then either 4 mm diameter pellets or dry mash to 48 days. Final live weight was 3.56 kg for pellets and 3.38 for mash. FCR was 2.26 for pellets and 2.20 for mash.

to achieve the same live weight as birds given pellets, it would be necessary to increase age at slaughter by about 2 days and then efficiency of feed conversion would be very similar to that of pellet-fed birds, but fixed costs would be higher.

In India, Central Asian Republics and the Far East, many producers avoid the energy costs involved in the manufacture and transport of mash and pellet feed, by supplying birds with locally grown cereal grains, supplemented with controlled quantities of concentrate rich in protein and other nutrients. Birds are also often encouraged to forage on rice paddy for a substantial part of their daily feed. Grain and concentrate can be fed *ad libitum* from hoppers from as early as 12 days, but must be supplied as a mixture in suitably designed waste-proof feeders (see Fig. 3.20), because birds prefer pellets. Given the opportunity, they will use their bills to rake feed out of the feeders to get at the pellets, wasting grain into the litter or the slurry if feeders are located over a slatted area.

Figure 5.34 describes a trial designed to investigate the effect of feeding different grain and concentrate mixtures of similar nutrient composition to compound pellet feed. Figure 5.35 shows the results. Gradually increasing whole wheat to 72% and 80% in grain and concentrate mixtures, respectively, reduced both daily feed intake and rate of gain. On two occasions feed was temporarily withdrawn and after about 3 h compound pellet-fed birds had no feed remaining in their oesophagus, but a substantial amount was palpated in birds given the grain and concentrate mixtures, suggesting the muscular activity

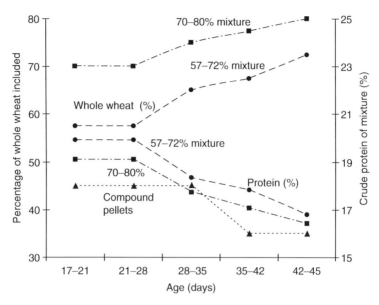

Fig. 5.34. Percentage of whole wheat included in mixtures of grain and protein concentrate in a trial designed to investigate the effect of increasing the percentage inclusion of whole grain with age. Birds were given pellet starter feed to 17 days, then grain and concentrate mixtures of similar energy (12.2 MJ ME/kg) to birds given compound pellet feed. The average trial temperature was 17°C.

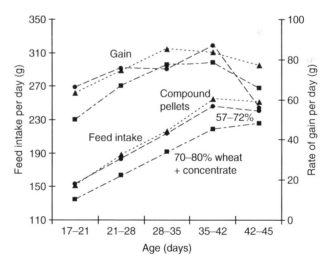

Fig. 5.35. Effect of increasing the percentage inclusion of whole grain in wheat and concentrate mixtures with age on feed intake and live weight gain. Percentage inclusion of whole wheat is shown in Fig. 5.34. Gradually increasing whole grain from 57% to 72% or from 70% to 80% reduced live weight at 45 days by 2.5% and 7.5% and increased FCR by about 1% and 2.5%, respectively.

required to grind wheat in the gizzard was sufficient to reduce feed motility and may explain why feeding whole grain and concentrate mixtures reduced feed intake.

Increasing environmental temperature from 10°C to 20°C depresses feed intake for birds given a wheat grain and concentrate mixture, compared to birds given pellets (Fig. 5.36). The compound pellets and the concentrate plus grain mixture were formulated to provide feeds of the same dietary energy and nutrient composition, but birds maintained at 20°C and given the grain and concentrate mixture were presumably forced to reduce their intake of feed to limit metabolic heat from muscular activity grinding grain, thus avoiding hyperthermia.

Factors Affecting Body Composition

Despite domestication over several thousand years, domestic ducks still share many similarities with their wild ancestors. Figure 5.37 describes the growth of Mallard and Pekin and shows a fivefold difference in absolute growth rate and mature body weight, but the pattern of growth and time to somatic maturity are

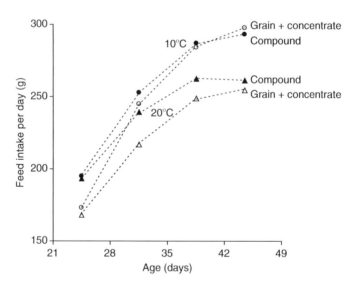

Fig. 5.36. Effect of temperature on feed intake for ducks given either a 75% wheat grain and 25% concentrate pellet mixture or compound pellets providing the same dietary energy and percentage of protein. All birds were housed intensively on litter in temperature-controlled accommodation and temperature was gradually adjusted to provide 10°C or 20°C when feed treatments commenced at 21 days. Providing a 75% wheat grain and concentrate mixture reduced live weight compared to birds given compound pellets by about 5% for birds reared at 20°C, but had no effect on feed intake or growth for birds grown at 10°C.

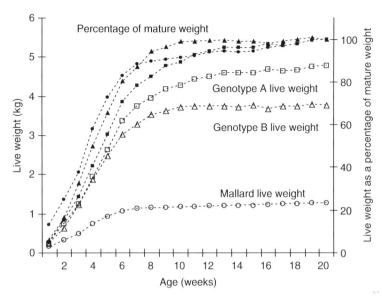

Fig. 5.37. Effect of age on growth of two Pekin genotypes (A) selected for growth rate and efficiency in feed conversion and (B) selected for breeding performance under controlled feeding conditions and game farm Mallard, measured in kilogram and expressed as a percentage of mature *ad libitum* live weight. (Data from Cherry, 1993.)

similar for wild Mallard and Pekin genotypes A and B, selected for growth rate and breeding performance, respectively.

Both Mallard and Pekin achieve mature body weight by about 12 weeks, when feathering and maturation of both primary and secondary wing feather is complete (see Fig. 5.40). Rapid early growth in Mallard is essential because the majority of wildfowl are hatched and reared in northern latitudes, and young birds must have enough breast muscle, wing area and body fat to allow them to migrate south with the adults to avoid low temperatures and food shortage during the northern winter.

In early spring, birds migrate north to breed, but poor supplies of available energy at that time of year, along with the effects of predation, mean that to successfully reproduce breeding females must have sufficient reserves of body fat to produce a clutch of about 12 eggs, and sustain them during incubation (Krapu, 1981). Breeding females lose about 20% of their body weight and total lipids decline from about 17% to 3% of body weight during migration, egg laying and incubation.

As soon as hatching is complete Mallard females leave their nesting site and do not return, leading chicks often a considerable distance to find a suitable habitat where they can find feed. Brooding Mallard females do not actively feed their chicks, and thigh and leg muscle and bone along with webbed feet grow rapidly to allow chicks to swim, feed and escape from predators. Figures 5.38 and 5.39 show modern genotypes of Pekin inherit this trait, achieving

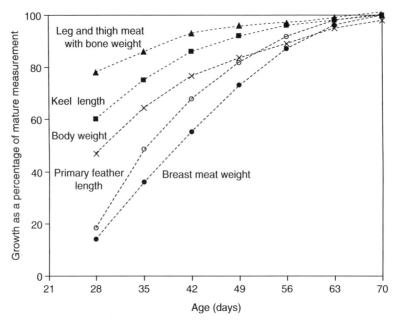

Fig. 5.38. Effect of age on growth of body weight, breast meat and thigh and leg meat (with bone), keel bone and sixth primary feather, expressed as a percentage of mature measurement. (Data from Cherry, 1993.)

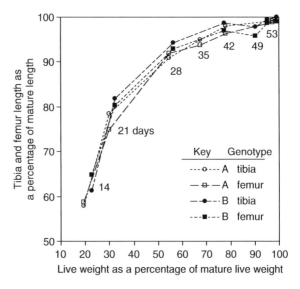

Fig. 5.39. Relationship between age, live weight and growth of tibia and femur length measured as a percentage of mature measurement for two genotypes with mature *ad libitum* live weights of 3.1 and 3.85 kg. (Data from M.S. Lilburn, Ohio State University, 2006, personal communication.)

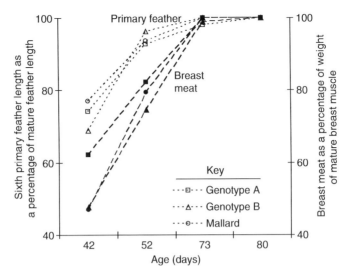

Fig. 5.40. Relationship between age and growth of breast meat and primary feather expressed as a percentage of mature measurement for Mallard, and Pekin genotypes A and B, with mature live weights of 1.25, 3.95 and 4.94 kg, respectively.

about 80% of mature measurement of leg bone and muscle as early as 28 days. Breast muscle and flight feathers grow rapidly from about 3 weeks of age (see Figs 5.38 and 5.40) to enable young Mallard to fledge by about 52 days (see Chapter 6, this volume), and allow birds hatching towards the end of July (from females producing second and third clutches of eggs to overcome the effects of predation) to migrate south with adults in the fall. The similarity in the pattern of growth of muscle for Mallard and Pekin reported by Gille and Salomon (1998), along with data presented in the next section of this chapter, shows that growth pathways and body composition of domestic duck are still profoundly affected by their genetic inheritance from wild Mallard.

Age, Sex, Maturity and Body Composition

Figure 5.38 describes the effect of age on growth to somatic maturity (measured as a percentage of mature measurement) and shows the pattern of growth for body weight, keel bone, breast meat and flight feathers, while Fig. 5.41 shows the effect of body weight on body composition for the same Pekin genotype selected for growth and feed conversion with a mature plucked body weight of about 5 kg. Analysis of the data provides an estimate for the effect of age (x, days) and body weight (y, g) on growth of breast meat expressed as a percentage and in grams.

Breast meat (%; for age 35–55 days) = $-26.439 + 1.2902x - 0.0113x^2$

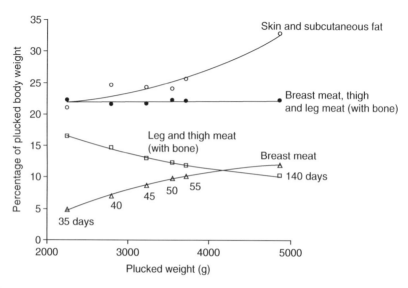

Fig. 5.41. Relationship between age and plucked body weight upon percentage of breast meat, leg and thigh meat (with bone) and skin and subcutaneous fat for slaughter ages from 35 to 140 days. Average of male and female Pekin genotype selected for growth and feed conversion with mature *ad libitum* live weight of about 5400 g.

Breast meat (g; for age 35–55 days) $= -1067.857 + 46.0429x$
$$- 0.3571x^2$$

Breast meat (%; for body weight 2200–3800 g) $= -3.405 + 0.0037y$

Breast meat (g; for body weight 2200–3800 g) $= -317.728 + 0.1868y$

Figures 5.42 and 5.43 describe the effect of genotype and age on growth of thigh and leg meat and of breast meat (both measured as a percentage of eviscerated weight). Figure 5.44 shows the growth of breast meat and keel bone mineralization (described as percentage of keel bone ash) for separate sexes. Analysis of these data provides the following estimates for the effect of age (x, days) between 40 and 49 days on percentage of keel ash.

Male bone ash (%) $= 122.57 - 5.95x + 0.0732x^2$

Female bone ash (%) $= 82.25 - 4.46x + 0.0617x^2$

Figure 5.45 shows a close curvilinear relationship between keel bone mineralization and proportion of breast meat for separate sexes slaughtered between 40 and 49 days. Both factors affect age at slaughter because cartilaginous ribs and keel collapse during roasting at high temperature and well-fleshed breasts are essential to satisfy customers. Both Figs 5.44 and 5.45 show that females mature earlier than males, as measured by percentage of both breast meat and bone ash, and so they can be slaughtered several days before males where there is a market for small but well-fleshed birds.

Figure 5.46 describes the relationship between age, body weight and breast meat for three Pekin genotypes. Genotypes A, B and C had about 150, 200

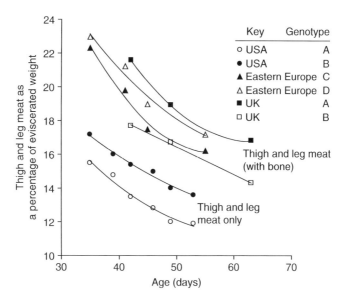

Fig. 5.42. Relationship between age and growth of leg and thigh meat for Pekin genotypes from the USA and leg and thigh meat (with bone) for Eastern Europe and UK genotypes recorded as a percentage of eviscerated weight. (Data from unreported trials and M.S Lilburn, Ohio State University, 2006, personal communication.)

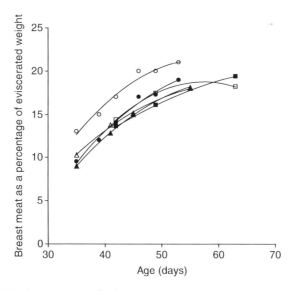

Fig. 5.43. Effect of age on growth of breast meat expressed as a percentage of eviscerated weight for Pekin genotypes from the USA, Eastern Europe and the UK. Data and key as in Fig. 5.42.

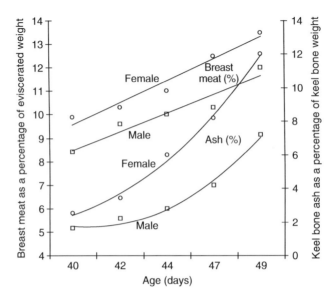

Fig. 5.44. Effect of age and sex on percentage of breast meat and keel bone ash for Pekin identified to sex and given starter to 17 days and a 12.55 MJ ME/kg and 17% protein diet to 49 days of age.

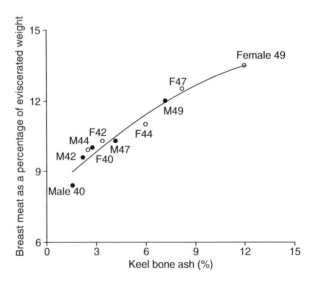

Fig. 5.45. Relationship between keel bone ash and breast meat recorded for males and females between 40 and 49 days. Average male and female percentage of ash was 3.6% and 6.4% and of breast meat was 10.1% and 11.5%, respectively. Diet as in Fig. 5.44.

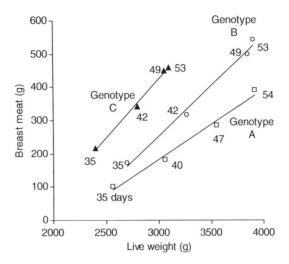

Fig. 5.46. The relationship between age, live weight and breast meat for Pekin genotypes selected for (A) growth and feed conversion, (B) growth and body composition and (C) body composition. Birds were reared in different locations but given diets of similar nutrient concentration. (Data from unreported trials and M.S Lilburn, Ohio State University, 1996, personal communication.)

and 400 g of breast meat, respectively, at the same body weight of 3000 g when about 39, 40 and 50 days old. Analysis of the data provides the following estimates for the effect of body weight (x, g) in the range 2600–4000 g for genotypes A and B, and in the range 2400–3000 g for genotype C, on breast meat between 36 and 53 or 54 days.

Genotype A breast meat (g) = −448.85 + 0.2103x

Genotype B breast meat (g) = −650.34 + 0.3014x

Genotype C breast meat (g) = −640.21 + 0.3548x

Figure 5.47 describes the same data, but with live weight measured as a percentage of 53- or 54-day live weight and breast meat measured as a percentage of live weight between 35 and 53 or 54 days. This shows that genotypes A, B and C, respectively, achieved about 65%, 70% and 78% of 53-day live weight by 35 days, and 4%, 6% and 9% breast meat measured as a percentage of 35-day live weight. Analysis of the data provides the following estimates for the effect of body weight (x, measured as a percentage of 53-day live weight) on breast meat measured as a percentage of live weight between 35 and 53 or 54 days.

Genotype A (%) breast meat = −7.3101 + 0.171x

Genotype B (%) breast meat = −9.9321 + 0.2353x

Genotype C (%) breast meat = −11.626 + 0.2656x

However, Fig. 5.48, which expresses both breast meat and live weight as a percentage of their weights at 53 or 54 days, shows that growth of breast meat

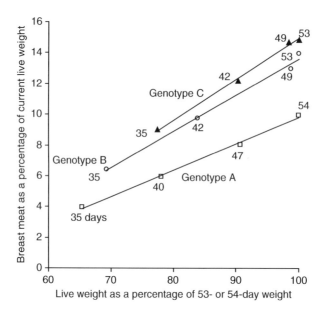

Fig. 5.47. Relationship between age, live weight as a percentage of 53- or 54-day live weight and breast meat as a percentage of the current live weight. Data source as in Fig. 5.46.

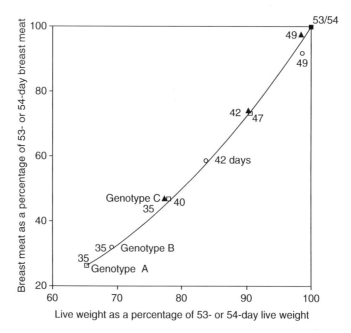

Fig. 5.48. The relationship between age and live weight (as a percentage of 53- or 54-day live weight) and breast meat (as a percentage of 53- or 54-day breast meat weight) for three Pekin genotypes. Data source and genotypes as in Fig. 5.46.

for all three genotypes was proportional to body weight. Analysis provides the following equation relating the growth of breast meat to body weight (x, percentage of body weight at 53 or 54 days).

Breast meat (percentage of 53- or 54-day breast weight)
$$= 55.487 - 2.1235x + 0.0257x^2$$

Analysis of trials reported in Chapter 6 (this volume) confirms that the principal factor affecting body composition for any given genotype is body weight because, in the trials reported, neither age, nor *ad libitum* or restricted feeding, nor alternative nutrient specifications had any substantial effect on body composition at a given weight (see Figs 6.29–6.32). The results of these trials along with the relationship between body weight and breast meat described in Fig. 5.48 confirm that the proportions of the Pekin duck appear to be determined entirely by relative growth. Body composition in the domestic duck appears to be a simple function of degree of maturity, as represented by body weight as a fraction of mature weight.

Market Requirements

The principal market outlets for duckling include Chinese restaurants, Michelin and premium-quality restaurants, cruise ships, fast-food restaurants and retail outlets selling whole and portioned duck, whole breast fillets and chilled oven-ready meals. These markets require different eviscerated weights and quality. Cruise ships, for example, need small birds suitable for portioning, but supermarkets and other retailers require large well-fleshed birds suitable for carving with sufficient meat to feed a family. Restaurants require premium-quality (200 g) breast fillets but producers of chilled oven-ready meals require small fillets of about 125 g. It is possible to use genotype, age at slaughter and nutrition to satisfy the wide and sometimes conflicting demands of these markets. The effect of genetic selection on growth and body composition is reviewed in Chapter 9 (this volume), and the effect of age at slaughter on eviscerated weight, body composition and processing yield is discussed towards the end of this chapter.

Nutrition and Body Composition

Figures 5.49–5.51 describe the effects of energy to protein ratio on body composition in unreported trials A and B and in experiments 1 and 2 from Scott and Dean (1991). Analysis of the data from these trials provides the following estimates for the effect of energy to protein ratio (x, kJ ME/g of protein) on various components measured as a percentage of eviscerated weight.

Breast meat (%) $= 15.777 - 0.0157x - 0.0005x^2$

Breast meat and leg meat with bone (%) $= 37.373 - 0.105x$

Breast meat and leg meat without bone (%) $= 23.441 + 0.096x - 0.0011x^2$

Skin and subcutaneous fat (%) $= 20.353 + 0.278x - 0.0004x^2$

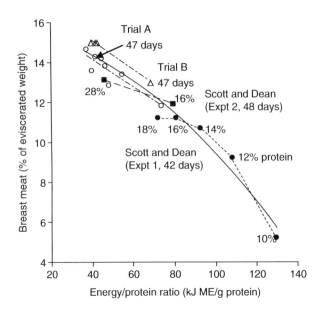

Fig. 5.49. Effect of energy to protein ratio upon growth of breast meat measured as a percentage of eviscerated weight at 42 and either 47 or 48 days of age. Data from unreported trials A and B where birds were given diets of 10.1, 11.3 and 12.6 MJ ME/kg and either 22% or 27% protein from 14 days of age, and from Experiments 1 and 2 of Scott and Dean (1991).

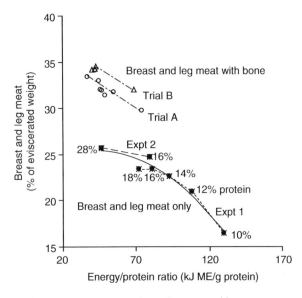

Fig. 5.50. Effect of energy to protein ratio on breast and leg meat expressed as a percentage of eviscerated weight recorded in four trials. Data sources as in Fig. 5.49.

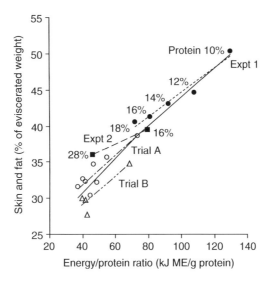

Fig. 5.51. Effect of energy to protein ratio upon skin and subcutaneous fat measured as a percentage of eviscerated weight between 42 and 48 days of age. Data sources as in Fig. 5.49.

Reducing the ratio of dietary energy to protein by 1 kJ/g of protein in the range 30–130 kJ/g of protein increased breast meat expressed as a percentage of eviscerated weight by about +0.09% and reduced skin and subcutaneous fat by about −0.2%. Large-scale commercial trials confirm that reducing the ratio from about 74 (12.6 MJ ME/kg and 17% protein) to about 37 kJ/g of protein (10.1 MJ ME/kg and 27% protein) increased breast meat by about 50–60 g and reduced subcutaneous fat by 120 g for birds with an average eviscerated weight of about 2 kg. However, feeding diets with narrow energy to protein ratios may affect eating quality, because taste panels and consumers in blind trials have described meat as firm, chewy and occasionally even tough. Birds having substantially less subcutaneous and intramuscular fat probably require cooking with techniques appropriate for Muscovy duck rather than those traditionally used for Pekin.

Factors Affecting Response to Energy to Protein Ratio

Figure 5.52 shows that reducing the energy to protein ratio and increasing age at slaughter from 47 to 53 or 54 days both substantially increased breast meat expressed as a percentage of eviscerated weight, although clearly the response would be affected by genotype (see Fig. 5.43). Results from commercial trials also confirm that increasing age at slaughter to about 54 days, the age when domestic ducks begin to moult body feathers, and feeding an energy to protein ratio of about 40 kJ/g of protein provides an opportunity to supply the market with relatively large but low-fat and well-fleshed birds, providing sufficient lean

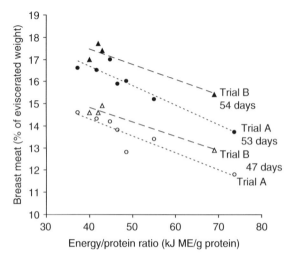

Fig. 5.52. Effect of energy to protein ratio on breast meat measured as a percentage of eviscerated weight for birds slaughtered at 47 and either 53 or 54 days in unreported trials A and B. Diets as in Fig. 5.49.

meat to feed a family. However, the opportunity to reduce age at slaughter to produce small but meaty birds is limited, because Fig. 5.53 shows that feeding a ratio of 37 kJ/g of protein diet increased the percentage of breast meat compared with a diet containing 73 kJ/g of protein, but did not increase bone mineralization as measured by the percentage of keel bone ash. Reducing the ratio of energy to protein improves lean yield but does not increase the rate of bone mineralization.

Figure 5.54 compares the effect of feeding energy to protein ratios of 75 and 57 kJ/g of protein on the body composition of two Pekin genotypes. It shows that reducing the ratio of energy to protein substantially reduced subcutaneous fat and increased the proportion of breast and leg meat for the selected genotype, but had a much smaller effect on the body composition of an unselected genotype with similar mature body weight.

Figure 5.55 presents the percentage of protein and the ratio of energy to protein in the diet selected by birds given a free choice of two feeds containing 12.6 MJ ME/kg and 13% or 32% protein, compared with a standard feeding programme. It is notable that the birds given a choice selected a much higher-protein diet than the controls at all stages. Analysis of the data provides the following estimates for the effect of age (x, days) on percentage of protein intake and the ratio of energy to protein selected by birds given a free choice.

Protein (%) = $21.76 + 0.40x - 0.0098x^2$

Energy to protein ratio (kJ/g of protein) = $58.86 - 1.02x + 0.0251x^2$

Figure 5.56 illustrates the effect of these two treatments on body composition and shows that allowing free choice of protein substantially increased breast and leg meat and reduced subcutaneous fat. The results of these trials along

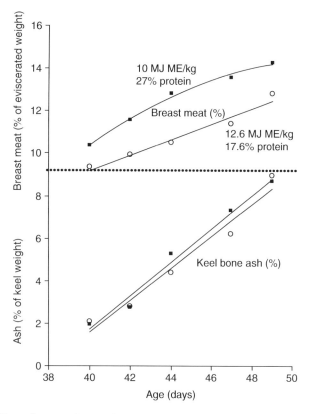

Fig. 5.53. Effect of age and ratio of energy to protein on breast meat measured as a percentage of eviscerated weight, and keel bone ash measured as a percentage of keel bone weight. Birds were reared intensively on litter and given starter to 17 days and then either 12.6 MJ ME/kg and 17.6% protein or 10.1 MJ ME/kg and 27% protein diets to 47 days of age, providing energy to protein ratios of 72 and 37 kJ/g of protein, respectively.

with evidence presented in Fig. 5.29 confirm that achieving genetic potential depends upon providing an adequate supply of energy and other nutrients, especially protein.

Response to energy to protein ratio can also be affected by environmental temperature. Figure 5.57 describes the body composition of birds of the selected genotype maintained at about 10°C or 20°C and given diets with energy to protein ratios of 71 or 53 kJ/g of protein. It shows that feeding more protein substantially reduced subcutaneous fat in the birds maintained at 20°C, but not at 10°C.

In large-scale trials investigating the effect of feeding a high-protein diet (10.25 MJ ME/kg and 27% protein; 38 kJ/g of protein) on commercial performance, birds excreted excess protein as uric acid and it was necessary to increase the amount of straw bedding provided daily to preserve feather quality and also substantially increase ventilation to maintain ammonia concentration at

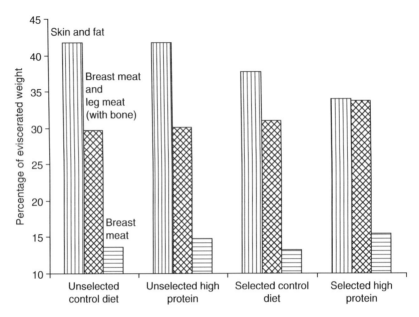

Fig. 5.54. Effect of dietary protein on body composition at 46 days for selected and unselected genotypes given starter to 17 days and then either 12.7 MJ ME/kg and 17% protein or 12.5 MJ ME/kg and 22% protein to 46 days, providing energy to protein ratios of 75 and 57 kJ/g of protein, respectively.

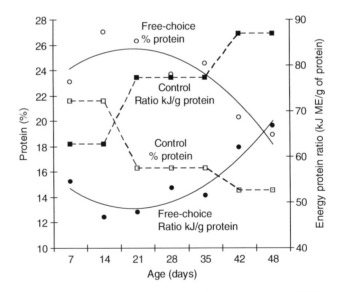

Fig. 5.55. Effect of age and providing a choice of feeds of 12.6 MJ ME/kg and 13% or 32% protein on percentage of protein intake and energy to protein ratio. Control birds were given starter to 14 days, then grower and fattener feeds of 12.6 MJ ME/kg and 16.3% and 14.5% protein to 35 and 48 days, respectively. Birds were reared intensively on litter. Choice feeds were provided from day-old. Live weights for control and birds fed with free choice at 48 days were 3.48 and 3.43 kg, respectively, and the feed conversion ratio (FCR) was 2.36 for both groups.

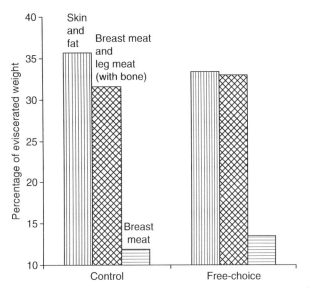

Fig. 5.56. Effect of providing birds with either a three-stage feed programme or a choice of feeds containing 12.6 MJ ME/kg and either 13% or 32% protein diets on body composition recorded at 47 days. Accommodation, diets and performance as in Fig. 5.55.

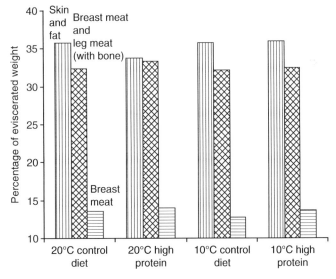

Fig. 5.57. Effect of temperature and dietary protein on body composition recorded at 48 days for birds of the same genotype maintained at 10°C or 20°C, given starter to 17 days and then 12.7 MJ ME/kg and either 18% or 24% protein to 48 days, providing energy to protein ratios of 71 and 53 kJ/g of protein, respectively.

less than about 10 ppm. Evisceration yield of birds given a ratio of 38 kJ/g of protein was about 3% lower than control birds, and reducing the ratio of energy to protein also reduced average live weight and subsequent oven-ready yield.

Although feeding extra protein reduces fat deposition and thereby increases the proportion of breast meat in the eviscerated carcass, it does not increase the absolute weight of breast meat at a given age. Underfeeding protein gives heavier carcasses with more subcutaneous fat and, in most markets, this strategy delivers a greater margin of profit.

References

Cherry, P. (1993) Sexual maturity in the domestic duck. PhD thesis, University of Reading, Reading, UK.

Cherry, P. and Morris, T.R. (2005) The maintenance requirement of domestic drakes. *British Poultry Science* 46, 725–727.

Fisher, C. and Wilson, B.J. (1974) Responses to dietary energy concentration by growing chickens. In: Morris, T.R. and Freeman, B.M. (eds) *Energy Requirements of Poultry*. Constable, Edinburgh, UK, pp. 151–184.

Gille, U. and Salomon, F.V. (1998) Muscle growth in wild and domestic ducks. *British Poultry Science* 39, 500–505.

Jung, Y. and Zhou, Y.P. (1980) The Pekin duck in China. *World Animal Review* 34, 11–14.

Krapu, G.L. (1981) The role of nutrient reserves in Mallard reproduction. *Auk* 98, 29–38.

Scott, M.L. and Dean, W.F. (1991) Energy, protein and amino acid requirements of ducks. In: Scott, M.L. and Dean, W.F. (eds) *Nutrition and Management of Ducks*. M.L. Scott of Ithaca, Ithaca, New York, p. 65.

Yunyan, Z., Wenlie, L., Lin, L. and Yongmei, R. (1988) Study on optimum ratio of dietary energy to protein for feeding meat-producing Beijing-duckling in South China. In: *Proceedings of the International Symposium on Waterfowl Production*. The Satellite Conference for the 18th World's Poultry Congress, Beijing, China. Pergamon Press, Oxford, pp. 193–199.

6 Rearing of Parent Stock

Until about 1970, breeding stock in the USA and Europe were selected from table ducklings fed for maximum live weight gain. Selection was based on growth rate and appearance, sometimes with an assessment of carcass quality made by handling the live bird. Selected birds were then reared to sexual maturity either extensively or in semi-intensive accommodation and with *ad libitum* feeding.

Choosing breeding stock in this manner exerted very effective selection for growth, because the number of birds retained for breeding was small in relation to the large number of ducks being produced and sent for processing. However, selecting for weight at a particular age also indirectly selected for increased appetite, fatness and mature weight and, unfortunately, for reduced reproductive fitness.

Figure 1.1 shows the effect of genetic selection on mature weight over the period from 1930 to the present time. Mature body weight increased substantially between 1960 and 1975 and anecdotal evidence suggests that by about 1970, egg production had declined from about 180 to about 133 eggs per female in a laying period of 45 weeks while hatchability had dropped from >70% to about 60%.

Faced with the problem of reduced reproductive performance, and knowing that the broiler industry had improved the reproductive potential of its breeding stock by restricting feed intake during growth, duck producers began to investigate restricted feeding.

Controlled Feeding

Figure 6.1 shows the effect of feed restriction on two Pekin genotypes. Genotype A was selected purely for growth and efficiency of feed conversion (mature weight 4.9 kg) and genotype B was selected for egg production as well as growth (mature body weight 4.1 kg). Both strains, when fed *ad libitum*,

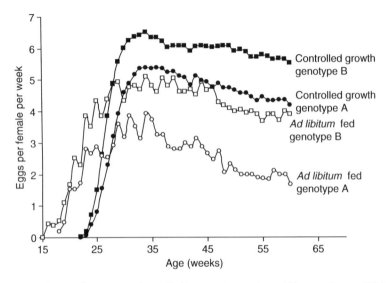

Fig. 6.1. Laying performance of two Pekin genotypes A and B reared on *ad libitum* and restricted feed to achieve 70% and 75%, respectively, of *ad libitum* mature weight at 18 weeks of age. To avoid precocious sexual maturity, both *ad libitum* fed and restricted birds were reared on a step-down non-stimulatory light programme of 23 h to 8 weeks, when daylength was decreased weekly in equal steps, to provide 17 h at 18 weeks of age. Birds were given *ad libitum* and timed feed restriction, respectively, from 21 weeks. (Data from Cherry, 1993.)

came into lay at about 16 weeks of age and reached 50% lay about a month earlier than the restricted birds. They produced about 100 and 170 eggs, respectively, by 60 weeks of age, whereas the restricted birds produced 173 and 211 eggs, respectively. There were differences between the two genotypes in the percentage of double-yolked eggs laid under *ad libitum* feeding but, in both strains, restricted feeding had a marked effect (Fig. 6.2). Increased mature size and *ad libitum* feeding to sexual maturity were both associated with increased multiple ovulation and lower egg production.

Controlling the food supply during rearing delays sexual maturity reduces multiple ovulations as measured by the incidence of double-yolked eggs and increases rate of lay for both egg-producing and meat genotypes of the Pekin breed.

In the Far East and China, parent stock are under strong natural selection for characteristics such as mobility and a robust capability to thrive in a particular environment, but are artificially selected to satisfy local market requirements in terms of egg size, colour, feather colour and down quality. Selection and rearing of parent and laying stock are still carried out in traditional ways. Most producers are marginal farmers who have no financial incentive to change a cost-effective method of rearing and selecting their stock. The vast majority of domesticated ducks in these countries continue to be reared extensively. However, population growth and an increasing standard of living in many Far Eastern countries have substantially increased local demand for table duck and

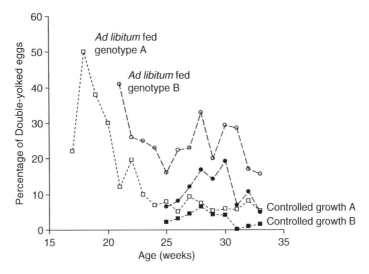

Fig. 6.2. Percentage of double-yolked eggs produced by two Pekin genotypes reared on *ad libitum* and restricted feed. Growth, lighting programme and laying performance as described in Fig. 6.1.

lower costs of production have encouraged the export of processed duck meat, feather and down to Western countries. This has led to the introduction of genotypes with better growth and feed efficiency from Europe and America along with semi-intensive and intensive methods of production. These new units also need to practise controlled rearing of their breeding stock.

Controlled Growth and Physiological Development

Commercial experience and the results of trials described in this and subsequent chapters confirm that controlling growth through feed restriction during rearing is critical to achieving maximum breeder performance. However, restricting growth also delays physiological development and this affects both environmental requirements and husbandry during rearing.

An immediate and highly visible effect of controlling growth by feed restriction from day-old is the slow development of down and feather of birds on feed restriction compared with full-fed birds. Down provides thermal insulation by trapping air around the body, thus reducing heat loss to the environment. The relatively slow growth of down of birds reared on feed restriction increases heat loss and delays the age at which birds can control their own body temperature.

A good method of assessing the effects of restriction on physiological development is to compare the growth of primary feathers. Figure 6.3 shows that controlling growth by feed restriction affected feather growth in a similar manner for both genotypes, and examination of body feathering confirmed that body feather growth was delayed, with juvenile down still visible on the

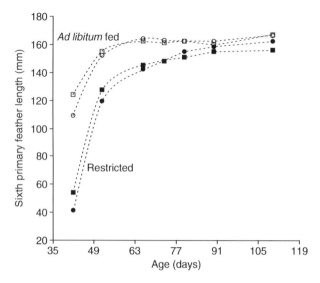

Fig. 6.3. Relationship between age and primary feather growth of Pekin females reared on *ad libitum* feed or restricted from day-old. Genotypes and growth as described in Fig. 6.1.

head and neck as late as 49 days. In *ad libitum* fed birds, the first juvenile moult of body feathers, sometimes referred to as 'second feather', normally commences at about 54 days, but this is delayed to about 84 days for birds reared on feed restriction.

Restricting growth improves mobility and vitality of growing ducks, partly due to a reduction in the prevalence of tibial dyschondroplasia. The breastbone of ducks normally calcifies over the period 5–8 weeks to form a substantial structure for breast muscle attachment but calcification is delayed until about 12 weeks in restricted birds and breast muscle growth, normally complete by about 8 weeks, is similarly delayed.

Table 6.1 lists the effect of *ad libitum* and restricted feeding on bone growth to various ages and shows that, while restriction reduces early bone growth, it has very little effect by 133 days of age. Another trial, in which live weight and shank length were recorded for two genotypes (Table 6.2), confirms that restriction to approximately 80% of *ad libitum* live weight had no permanent stunting effect on bone growth.

Normally, ducks develop the ability to quack and drakes produce a characteristic hiss by about 35 days and this difference is sufficient to enable accurate segregation of sexes at that age. However, when feed is restricted, the ability to quack and the characteristic difference between sexes does not become apparent until about 56 days.

Ducks reared on *ad libitum* feed and provided with a constant daylength of 17 h exhibit sexual display behaviour, with females approaching males and dipping their heads in a characteristic manner approximately 5–6 weeks before first egg. Observation of birds on feed restriction provided with the same light-

Table 6.1. Effect of *ad libitum* and restricted feed upon shank length in a meat strain of Pekin duck. The restricted ducks reached 58% of the weight of *ad libitum* fed ducks at 50 days and 80% at 133 days. (Data from Cherry, 1993.)

	Shank length (mm)		Control-fed as a
Age (days)	*Ad libitum* fed	Control-fed	percentage of *ad libitum*
50	54.8	41.3	75
91	53.1	46.7	88
133	52.3	50.7	97

Table 6.2. Effect of *ad libitum* and restricted feed upon live weight and shank length at 105 days of age in two Pekin genotypes. (Data from Cherry, 1993.)

Genotype	A		B	
Feed programme	*Ad libitum*	Restricted	*Ad libitum*	Restricted
Live weight (kg)	4.31	3.28	4.18	2.95
Shank length (mm)	55	59.8	56.0	53.3

ing programme indicates that display behaviour is delayed by several weeks, but occurs at about the same interval before first egg.

Table 6.3 presents data from an experiment in which two genotypes were reared on *ad libitum* and restricted feed and samples of birds were examined for ovary and oviduct development when each group reached 50% lay and again at the end of a laying cycle of 40 weeks. Growth in the restricted birds was approximately 80% of that of the *ad libitum* fed groups. Internal ovulation was relatively frequent in all groups. The principal clinical symptom of this condition was the presence of variable amounts of yolk material being resorbed within the body cavity. Restricted birds actually showed a higher incidence of internal laying than full-fed birds at point of lay, but this ranking was reversed by the end of the laying cycle. Internal ovulation is likely to occur whenever the organization of follicle development and growth is not properly synchronized, and therefore is more prevalent when birds are approaching sexual maturity and the hormone regulation system controlling ovulation has not properly settled down.

Rearing birds on *ad libitum* feed increased the number of ripe follicles (>5 mm) present at sexual maturity compared to birds reared on feed restriction, which supports the evidence in Fig. 6.2 that *ad libitum* fed birds produced significantly more double-yolked eggs. However, at the end of lay the number of rapidly developing follicles was greater in the birds reared on restriction (Table 6.3).

Several birds also showed clinical evidence of follicular atresia, which appeared to be of two types. The first, characterized by an irregular and discoloured surface to the follicle, has been described in domestic fowl by

Table 6.3. Effect of *ad libitum* and restricted feeding during rearing upon number of birds showing clinical signs of internal laying, number of ripe ova (>5 mm) in the ovary and mean oviduct weight. From the same trial as in Table 6.2.

Genotype	Rearing treatment	Stage of lay	Number of females	Number with symptoms of internal laying	Percentage	Number of ova >5mm	Weight of oviduct (g)
B	*Ad libitum*	50%	18	7	39	11.0	86.2
	Restricted		9	5	55	9.7	87.8
	Ad libitum	End of lay	10	7	70	5.6	91.2
	Restricted		12	6	50	6.4	78.3
A	*Ad libitum*	50%	22	6	27	12.8	97.4
	Restricted		11	7	64	7.8	100.0
	Ad libitum	End of lay	9	8	89	5.9	93.4
	Restricted		12	4	33	7.3	62.8

Gilbert (1979) as 'bursting atresia', and the second, by a marked reduction in size and discoloration of the follicle, suggesting resorption, described as 'invasion atresia'. This condition has previously been reported in the domestic duck by Chueng-Shyang Ma (1968).

In summary, feed restriction delays growth and physiological development as measured by the effect of age on bone and feather growth, behaviour and sexual development. These effects are quantitatively related to the degree of restriction, measured as a percentage of the *ad libitum* live weight. Examination at sexual maturity indicates that restricting growth to approximately 80% of normal growth reduces the incidence of, but does not prevent, the relatively abnormal follicle growth recorded in ducks of heavy genotypes reared with *ad libitum* feeding.

Effect of Controlled Nutrient Intake on Growth

Cherry (1993) carried out several trials investigating the effects of controlled nutrient intake on growth and subsequent breeding performance. The first trial, which investigated the effect of energy and protein on early growth to 56 days, is illustrated in Fig. 6.4. Figure 6.5 shows the effect of energy restriction, measured as total energy intake to 56 days, upon live weight attained in this trial.

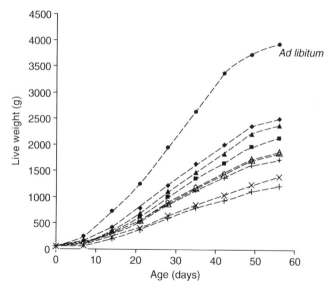

Fig. 6.4. Effect of feed restriction on growth to 56 days of Pekin genotype B. Eight levels of feed restriction were provided by using three diet specifications, which with suitable levels of feed restriction provided progressive reductions in nutrient concentration. Feed restriction commenced at day-old and continued to 56 days; growth was controlled within the range 30–65% of *ad libitum* growth. (Data from Cherry, 1993.)

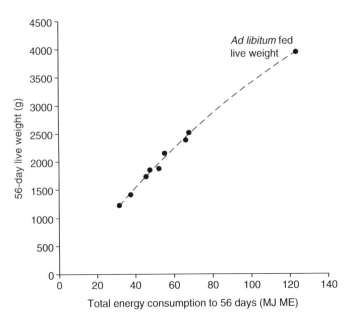

Fig. 6.5. Effect of energy restriction, measured as total energy consumption to 56 days, upon live weight. Treatments and genotype as in Fig. 6.4. (Data from Cherry, 1993.)

Figure 6.6 shows the relationship between lysine consumption and growth to 56 days. There are several treatments with large differences in lysine intake but similar live weight, indicating that lysine was not limiting growth in this trial. Taken together, Figs 6.5 and 6.6 indicate that the factor controlling growth in these treatments was energy intake.

These results show the large extent to which growth can be controlled by limiting feed intake but there are practical considerations which impose limitations on the extent to which feed restriction can be practised. Providing <90 g of feed per duck per day increases competition to an extent that produces inequitable feed consumption. Under commercial conditions this will increase variation in live weight which could affect subsequent breeding performance.

In three further large-scale trials, growth was controlled to give live weight at 20 weeks in a range from 2.5 to 4 kg, which is about 50–80% of *ad libitum* weight. This was achieved by restricting the intake of diets with energy contents varying from 10.25 to 13.52 MJ ME/kg. Analysis of results of these trials provides estimates of the effect of restricted energy consumption (x, MJ ME) on growth of Pekin females to 8 and 20 weeks of age:

8-week live weight (kg) = $0.416 + 0.02803x$

20-week live weight (kg) = $0.503 + 0.01302x$

The results of these trials further demonstrate that, provided protein is not made limiting, growth to 8 weeks and also to 20 weeks of age is a function of energy restriction (see Figs 6.7 and 6.8).

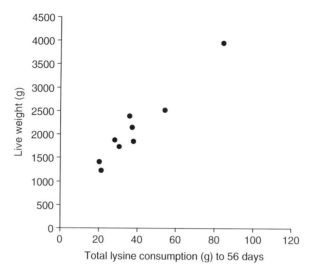

Fig. 6.6. Relationship between cumulative lysine consumption and growth of Pekin females. Treatments and genotype as in Fig. 6.4. (Data from Cherry, 1993.)

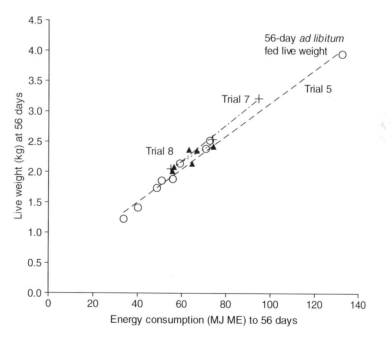

Fig. 6.7. Relationship between cumulative energy consumption and live weight at 56 days for Pekin genotype A females. Growth in three trials was controlled in the range 30–80% of *ad libitum* 56-day live weight, by restricting feed intake of diets with energy contents varying from 10.25 to 13.52 MJ ME/kg. (Data from Cherry, 1993.)

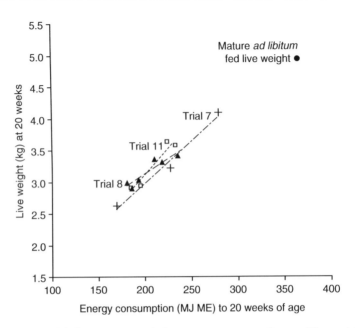

Fig. 6.8. Relationship between cumulative energy consumption and live weight of Pekin genotype A females. Growth to 20 weeks in three trials was controlled by feed restriction, using diets with similar energy contents to those described in Fig. 6.7 to control growth in the range 50–80% of *ad libitum* mature live weight. (Data from Cherry, 1993.)

Effect of Age upon Potential for Growth

Figure 6.9 shows the results of an experiment in which groups of females were grown with feed restriction to 7, 12, 17 or 22 weeks and then given *ad libitum* feed. Growth and feed intake were compared with a control treatment of the same genotype reared on *ad libitum* feed to 32 weeks of age. Ducks restricted from day-old but subsequently allowed access to *ad libitum* feed at 7 or 12 weeks increased their feed consumption to an extent that allowed them to compensate for the early restriction and achieve a similar mature weight to birds given *ad libitum* feeding throughout. In contrast, birds restricted until 17 or 22 weeks did not grow and did not increase their feed intake when provided with *ad libitum* feed. Consequently, their mature weight was approximately 85% of the other treatments at 32 weeks of age.

The ducks in the above trial were given 17 h of light a day from 16 weeks onwards. However, in a subsequent trial, where ducks were reared on a short daylength of 8 h until either 18 or 24 weeks, quite different results were obtained. The birds were grown to 50%, 65% or 80% of *ad libitum* live weight until 28 weeks of age, when additional feed was supplied to the 50% and 65% groups (see Fig. 6.10). In this case, the severely restricted birds responded to their increased feed supply and reached the same live weight as the 80% group by 34 weeks of age. The explanation for the different effects of age upon

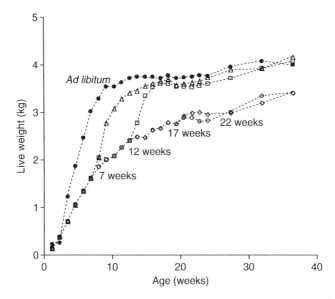

Fig. 6.9. Relationship between age and potential for growth of Pekin genotype B females, *ad libitum* mature weight 4.1 kg. Birds were reared on feed restriction to 7, 12, 17 or 22 weeks and then allowed *ad libitum* feed. They were given a constant daylength of 23 h to 7 weeks after which daylength was reduced weekly to reach 17 h by 16 weeks of age. (Data from Cherry, 1993.)

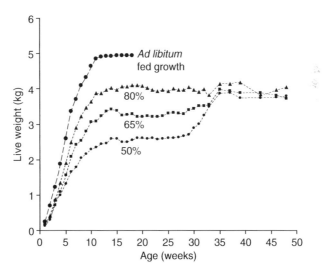

Fig. 6.10. Growth of Pekin genotype A females reared on feed restriction to reach 50%, 65% or 80% of *ad libitum* mature live weight at 28 weeks of age, when the daily feed allowance was increased for the 50% and 65% treatments. Timed feed restriction was provided to all treatments from 34 weeks of age. Birds were given a daylength of 23 h to 8 weeks, then provided with an 8 h day to either 18 or 24 weeks, when daylength was increased by 1 h per week to 17 h. (Data from Cherry, 1993.)

capacity for growth shown in these and other trials reported later in this chapter, where females showed compensatory growth after 17 weeks, appears to be that growth can potentially continue until development of the ovary and its increasing hormone production inhibits or diverts the effect of growth hormones. As shown later in this chapter, light, in addition to live weight, significantly affects age at sexual maturity. Figure 6.9 thus describes the effect of age on potential for growth of birds reared under long days throughout, whereas Fig. 6.10 shows the response of birds reared from 8 weeks on a daylength of 8h. Short days during rearing delay sexual development and so allow growth to continue well beyond 17 weeks of age. Catch-up growth of the skeleton is possible before sexual maturity, but not afterwards.

Effect of Temperature on Growth to Sexual Maturity

The effects of temperature on full-fed ducklings up to 49 days of age were discussed in Chapter 3 (this volume). There is very little information about the effects of temperature on the growth of parent stock to sexual maturity, but one experiment investigated the effects of high ambient temperature upon growth and sexual maturity of birds reared simultaneously on *ad libitum* and restricted feed in the UK and in Singapore where temperature and humidity remain high throughout the year.

Females of the same genotype, provided with similar light programmes, were reared to 20 weeks of age on *ad libitum* feed at two temperatures in the UK (11°C and 26°C) and in Singapore where the mean temperature was 27°C. Other groups were reared on restricted feeding in Singapore and at the lower temperature in the UK. Figure 6.11 shows the effect of temperature and *ad libitum* and restricted feed on growth to 20 weeks of age at both locations.

The effect of high temperature on *ad libitum* fed birds was to reduce live weight at 20 weeks at both locations, to about 83% in the UK and 77% in Singapore of the weight of ducks grown on *ad libitum* feed at 11°C. However, this did not affect physiological development as measured by growth of primary feathers in the UK (see Fig. 6.12) but did delay age at 50% lay by about 10 and 17 days relative to birds reared on *ad libitum* feed at the lower temperature of 11°C (see Table 6.4). Temperature at both locations was measured as dry bulb temperature but there were substantial differences in relative humidity and radiant temperature between the locations affecting 'effective temperature'. This is probably responsible for the lower live weight at 20 weeks and later sexual maturity of the ducks in Singapore.

The evidence from this trial indicates that high ambient temperature reduces growth rate and mature live weight, but produces a smaller delay in physiological development, as measured by feather growth and age at sexual maturity, than feed restriction at normal temperature applied to give a comparable reduction in live weight.

The controlled growth treatments at both locations were reared on the same controlled growth curve to 20 weeks of age (see Fig. 6.11), but on quite

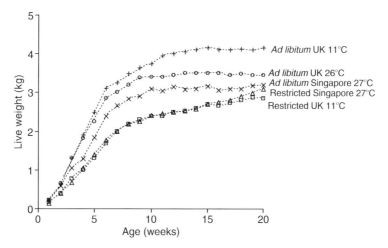

Fig. 6.11. Growth of Pekin genotype B females reared at 26°C and 11°C in the UK and at 27°C in Singapore. Birds were reared either on *ad libitum* feed or restricted (at both locations) to achieve 75% of UK *ad libitum* mature live weight at 20 weeks. All birds were given a step-down lighting programme of 23 h to 5 weeks when daylength was reduced weekly in equal steps to provide 17 h by 17 weeks of age. The 26°C treatment in the UK was maintained by indirect gas air heating. (Data from Cherry, 1993.)

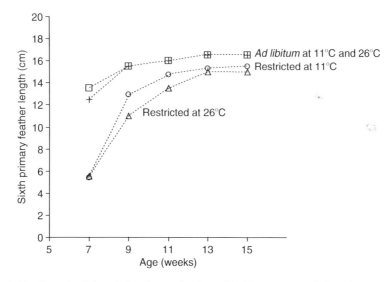

Fig. 6.12. Growth of the sixth primary feather of Pekin genotype B females reared on *ad libitum* and restricted feed at 11°C and 26°C. Treatments and genotype as in Fig. 6.11. (Data from Cherry, 1993.)

different growth curves when measured as a percentage of the local *ad libitum* growth curves. Females reared in Singapore were approximately 70% of UK 11°C *ad libitum* live weight, but almost 90% of Singapore *ad libitum* live weight.

Table 6.4. Effect of ambient temperature at two locations on weight and sexual maturity of *ad libitum* fed and restricted Pekin ducks. (Data from Cherry, 1993.)

Treatment	Temperature (°C)	Live weight (kg) at 20 weeks	Live weight as a percentage of treatment 1	Age (weeks) at sexual maturity
1. UK, *ad libitum*	11	4.15	100	19.7
2. UK, restricted	11	3.10	75	25.5
3. UK, *ad libitum*	26	3.45	83	21.0
4. Singapore, *ad libitum*	27	3.20	77	22.0
5. Singapore, restricted	27	2.86	69	24.0

Effect of Restricted Growth on Age at Sexual Maturity

Olver (1988) investigated the effects of restricting feed intake to 50% of *ad libitum* from 3 or 8 weeks to 20 weeks of age upon sexual maturity and breeding performance of Pekin females having an *ad libitum* mature live weight of about 4.1 kg. He showed a linear regression of age at sexual maturity (y, days) on relative live weight at 20 weeks (w = percentage of *ad libitum* live weight):

$$y = 256.9 - 1.03w$$

There was also a linear effect of relative weight upon laying performance:

Eggs per female to 60 weeks of age $= 179.6 - 0.353w$

Olver *et al.* (1978) had previously examined the effects of restricting feed intake from 8 weeks to 75% and 50% of *ad libitum* feed intake, and in another trial to 80%, 60% and 40% of *ad libitum* feed intake, upon growth to 22 weeks using local South African strains of Pekin. Analysis of data from both trials shows that increasing severity of feed restriction from 8 weeks reduced growth to 22 weeks and increased subsequent age at sexual maturity in a linear manner. Live weight at 22 weeks (w) expressed as a percentage of *ad libitum* live weight was represented by:

$w = 35.96 + 0.696f$ (where f = percentage of *ad libitum* feed intake)

Age at sexual maturity was given by y (days) $= 267.5 - 0.749w$

There was no significant effect upon laying performance. The difference in the reported effect of live weight upon age at sexual maturity may well be a consequence of the different genotypes and perhaps lighting programmes used during rearing, which were not reported for either experiment. They reported that feed restriction reduced overall feeding costs, and concluded that the most profitable level of restriction was between 50% and 60% of *ad libitum* feed intake during the rearing period.

Olver (1984a,b) also compared the effects of restricting feed intake from 8 weeks (60% of *ad libitum* feed intake) with feeding a low-lysine (0.35%) diet *ad libitum*. Only quantitative feed restriction was effective in delaying sexual maturity and no significant difference was found between treatments for egg number, egg weight, fertility or hatchability. However, the fact that restricted birds were delayed in sexual maturity by 24 days shows that feed restriction had a significant effect upon intensity of lay since they achieved the same egg numbers per female as the *ad libitum* fed ducks.

Olver (1986) reported three further trials investigating the effects of feeding low-lysine diets *ad libitum* from various ages upon sexual maturity and subsequent breeding performance and concluded that only feeding low lysine from day-old was effective in restricting growth to 20 weeks of age and delaying age at sexual maturity.

Analysis of all seven trials reported by Olver *et al.* between 1978 and 1988, with differences between trials removed by least squares, provides an estimate of the effect of live weight, measured at 20–22 weeks, upon age at sexual maturity:

$$\text{Age at sexual maturity (days)} = 250.0 - 0.641w + k_i$$

where w = live weight expressed as a percentage of *ad libitum* weight and k_i is an adjustment for the *i*th trial.

The seven trials show a consistent effect of live weight upon age at sexual maturity, with each 10% reduction in live weight associated with 6.4 days delay in maturity, but indicate little if any advantage in terms of overall laying performance. However, those treatments which delayed maturity by 10–20 days did not, in general, reduce the number of eggs laid to a fixed age.

Hocking (1990) investigated the effects of feed restriction during rearing on ovarian function at the onset of lay in Pekin females similar to genotypes A and B. He concluded that there was a linear relationship between the number of yellow follicles in the ovary and body weight at onset of lay, which suggested a direct link between growth and ovulation rate. He suggested that the optimum degree of restriction was close to the minimum body weight required to achieve the sexual maturity recorded in his trials, which was 0.6 of the unrestricted weight for both genotypes. This is based upon the fact that the optimum number of yellow follicles, which he suggested was six, was observed in birds at or near this proportion of *ad libitum* weight. Since the majority of ducks of both genotypes reared at 0.5 of *ad libitum* body weight had failed to lay by 35 weeks, he suggested that a minimum weight may be required to initiate ovulation.

Effect of Controlled Growth on Laying Performance

Cherry (1993) reported the results of several trials designed to investigate the effects of growth and mature live weight (expressed as a percentage of *ad libitum* live weight) upon age at sexual maturity and subsequent breeding performance.

A large-scale well-replicated factorial trial investigated the effect on sexual maturity and subsequent laying performance of limiting the growth of Pekin genotype A (*ad libitum* mature weight 4.94 kg) to 50%, 65% or 80% of *ad libitum*

148

Chapter 6

growth. Feed restriction commenced at day-old. Live weight was measured weekly and feed restriction was adjusted as necessary to maintain the three treatments upon the intended growth curve to maturity (see Fig. 6.10). A step-down, step-up light programme was provided; all birds were reared on 23h daylength to 7 weeks when daylength was abruptly reduced to 8h until either 18 or 24 weeks, after which daylength was increased in increments of 1h per week to 17h.

All growing treatments were maintained at 50%, 65% and 80% of *ad libitum* live weight through continued restriction until about 28 weeks, when it became clear that birds maintained at 50% and with increase in daylength at 18 weeks were not coming into lay in a normal manner (see Fig. 6.13) and had started to moult. Daily feed allowance was then increased to birds maintained at 50% and 65% of mature weight and Fig. 6.10 shows there was an immediate increase in live weight. From about 34 weeks of age all birds were given 5h feeding time per day, and by 37 weeks all restricted treatments weighed approximately 80% of *ad libitum* mature live weight. This was effective in preventing the other 50% treatment, which received increased daylength from 24 weeks, from moulting. Live weight at 20–30 weeks significantly affected age at sexual maturity, and the results indicate that there is a minimum threshold live weight between 50% and 65% of *ad libitum* live weight which is essential for achieving sexual maturity and continued egg production.

Another trial investigated the effects of rearing and maintaining the same Pekin genotype at 75% and 65% of *ad libitum* mature live weight. Birds were

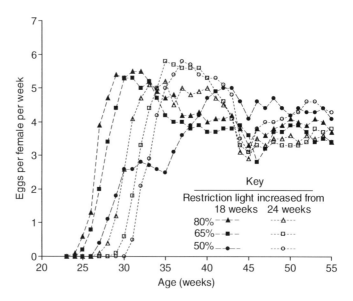

Fig. 6.13. Laying performance of Pekin genotype A reared and maintained at 50%, 65% and 80% of *ad libitum* weight until 28 weeks, when feed allowance was increased to birds on the 50% and 65% treatments. From 35 weeks, all birds were given 5.5h of restricted feeding time. Birds were reared on a step-down, step-up lighting programme with increased daylength from 18 or 24 weeks. (Data from Cherry, 1993.)

reared on feed restriction from day-old to 29 weeks (see Fig. 6.14), but in order to maintain the lower live weight treatment at the intended percentage of *ad libitum* live weight in lay, it was necessary to severely restrict feed consumption in contrast to the other heavier growing treatment, which was given timed feed restriction of 5.5h daily from 29 weeks of age. Rearing and maintaining birds at 75% and 65% of mature weight delayed sexual maturity to about 28 and 33 weeks of age, respectively (see Fig. 6.15), but maintaining birds at 65% of *ad libitum* mature live weight reduced rate of lay. At 34 weeks of age, their daily feed allowance was increased by about 20% to 190g per bird per day, which substantially improved rate of lay, but also allowed birds to gain weight. By 50 weeks of age, birds on the two treatments weighed 79% and 73% of *ad libitum* mature live weight, respectively.

Both trials show that restricting growth is effective in delaying sexual maturity, but the extent to which it can be delayed is limited by the fact that growth of <60% of *ad libitum* mature live weight prevents normal development of the ovary and subsequent sexual maturity. Controlling feed intake in lay to hold live weight to <70% of *ad libitum* live weight reduces laying performance.

In a further trial, as part of a factorial experiment investigating the effect of body composition upon age at sexual maturity, genotype A females were reared on feed restriction and weighed 64% and 73% of *ad libitum* mature weight at 26 weeks of age when timed feed restriction was introduced. Birds were given 2h feeding time per day, increased by 1h a week, to a maximum of 6h, and by 29 weeks both treatments were similar in live weight, and by 34 weeks birds weighed 75% and 76%, respectively, of mature live weight.

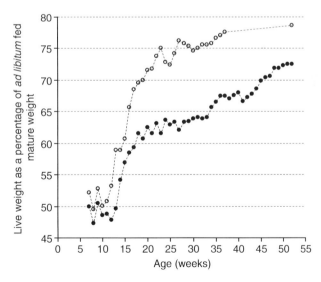

Fig. 6.14. Growth of genotype A reared to achieve 65% and 75% of *ad libitum* live weight of 4.94kg by 20 weeks and then maintained at this percentage of mature weight until 29 weeks by continued quantitative feed restriction. Birds were given a step-down lighting programme providing 23h to 8 weeks, when daylength was reduced weekly in equal steps to provide 17h at 18 weeks and during lay. (Data from Cherry, 1993.)

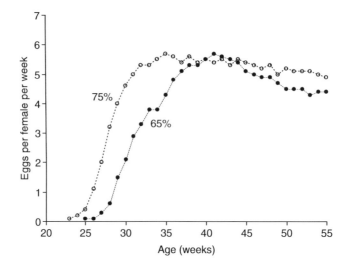

Fig. 6.15. Laying performance of Pekin genotype A maintained at about 65% and 75% of *ad libitum* mature live weight by continued feed restriction in lay. Controlled growth, feed restriction and lighting programme as in Fig. 6.14. Birds reared to achieve, and then maintained at, 75% of mature weight in lay were provided with timed feed restriction of 5.5 h from 29 weeks. Quantitative feed restriction was required to maintain birds at 65% of mature weight in lay and this adversely affected rate of lay, making it necessary to increase feed by about 20% from 34 weeks. Egg production to 55 weeks of age was 148 and 122 eggs, respectively, per bird housed. (Data from Cherry, 1993.)

The evidence from these and other trials reported in this chapter suggests that regardless of the degree of restriction in growth to onset of lay, as soon as the feed supply is increased to allow normal egg production, live weight increases to >70% of *ad libitum* mature live weight. Figure 6.16 describes the increase in live weight from 18 to about 50 weeks of age in four trials using genotype A females grown to achieve between 50% and 80% of mature weight at 26 weeks of age. The evidence suggests that mature weight was not affected by growth to 18 or 26 weeks, age at sexual maturity or age when birds were allowed access to feed for >5 h per day.

To achieve sexual maturity and sustain a normal rate of lay birds must achieve >70% of *ad libitum* mature live weight. There appears to be little to be gained from the widespread commercial practice of continuing quantitative feed restriction after about 18 weeks of age, to prevent birds gaining weight when approaching sexual maturity.

Pattern of Growth and Age at Sexual Maturity

As part of the same series of trials investigating the effect of controlled growth on age at sexual maturity, Pekin genotypes A and B were grown to achieve 75% of their *ad libitum* mature weight at either 18 or 26 weeks of age.

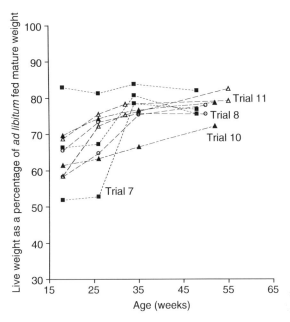

Fig. 6.16. Increase in live weight during lay expressed as a percentage of *ad libitum* mature weight. Controlled growth treatments in four trials using a Pekin genotype with a mature live weight 4.94 kg. Birds were reared to achieve between 50% and 80% of *ad libitum* mature live weight at 26 weeks of age. Birds in the four trials received different light programmes and all, with the exception of one treatment in trial 10, were provided with timed feed restriction of at least 5.5 h. (Data from Cherry, 1993.)

Figure 6.17 describes their relative growth to 32 weeks of age. Birds were given a step-down lighting programme and provided with timed feed restriction of 2 h per day from 26 weeks of age, increasing by 1 h a week, to a maximum of 5 h feeding time.

Figure 6.18 shows the effect of these two treatments on subsequent laying performance to 55 weeks. Genotypes A and B grown to 75% of mature weight at 18 weeks achieved sexual maturity at 27 and 26 weeks, respectively, 2 and 3 weeks earlier than the slower growing treatments, and this increased egg numbers per bird to a fixed age of 55 weeks by 12.7 and 7.1 eggs, respectively.

In a similar large-scale trial, genotypes A and B were reared to achieve 70% and 80% of mature weight at either 18 or 26 weeks of age. Birds were given a step-down lighting programme and all treatments were reared on quantitative feed restriction until birds achieved 10% lay, when they were provided with timed feed restriction of 3 h, increasing by 1 h a week to a maximum of 6 h. Results were similar to the trial shown in Fig. 6.18. Genotypes A and B grown to 70% of *ad libitum* live weight at 18 weeks achieved sexual maturity at 27 and 25.5 weeks of age, 5 and 1.5 weeks before the slower growing treatments, and produced 143 and 178 eggs per bird, 20 and 2 eggs more, respectively, to a fixed age of 55 weeks than the slower growing treatments.

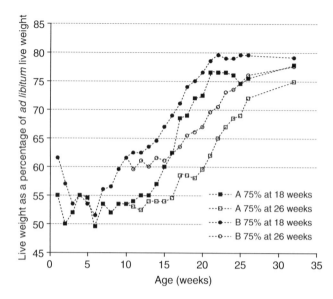

Fig. 6.17. Growth to 32 weeks of Pekin genotypes A and B grown to achieve 75% of *ad libitum* mature live weight of 4.94 and 4.1 kg at 18 or 26 weeks of age. Birds were reared on a step-down light programme providing 23 h to 8 weeks when daylength was reduced weekly in equal steps, to provide 17 h at 18 weeks and during lay. (Data from Cherry, 1993.)

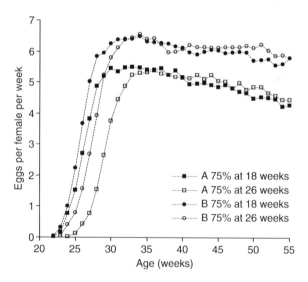

Fig. 6.18. Laying performance of two Pekin genotypes A and B with *ad libitum* mature weights, controlled growth and light programme during rearing as in Fig. 6.17. Two hours restricted feed time was provided from 26 weeks, increased by 1 h each week to 5 h. (Data from Cherry, 1993.)

These two trials show that allowing birds to achieve 70–75% of *ad libitum* mature live weight by 18 weeks reduces age at sexual maturity and increases laying performance to a fixed age when compared with programmes supporting lower growth rates.

The Importance of Early Feed Restriction

Commercially, early feed restriction is considered absolutely essential to achieve optimum performance over and above any effects on mature live weight and sexual maturity. However, improvement in laying performance from birds reared on restriction, compared with birds reared on *ad libitum* feed, could be attributed to reduced mature live weight, delayed sexual maturity and increased intensity of lay, or it could be a consequence of some other factor associated with feed restriction. This latter hypothesis was first proposed by Hollands and Gowe (1961), who suggested that feed restriction might act as a mild stress to stimulate greater development of the endocrine glands. This idea is supported by the evidence of improved laying performance of restricted birds in second-year production (see Chapter 7, this volume), which seems to indicate a permanent effect of restriction upon laying performance.

To investigate the effects of early feed restriction without a reduction in mature size on age at sexual maturity and subsequent laying performance, genotype B females were reared on feed restriction to 9 weeks and then allowed to achieve *ad libitum* live weight prior to sexual maturity. Their growth (see Fig. 6.19), age at sexual maturity and laying performance (see Fig. 6.20) were then compared with birds reared on restriction to achieve 75% of *ad libitum* mature live weight by 18 weeks. Early restriction to 9 weeks followed by *ad libitum* feeding delayed sexual maturity by approximately 2 weeks, compared with the same genotype reared on *ad libitum* feed from day-old (see Fig. 6.1), and very substantially improved rate of lay to 40 weeks of age.

Birds provided with *ad libitum* feed from 9 weeks reached 50% lay at 24.5 weeks and laid 179 eggs per bird to 55 weeks, compared with figures of 28 weeks and 164 eggs for the group reared to 75% of *ad libitum* weight at 18 weeks. Given the large difference in live weight between treatments in this trial, it is clear that live weight at sexual maturity and during early lay has only a small effect on subsequent rate of lay. Rearing birds on *ad libitum* feed from 9 weeks reduced age at sexual maturity and significantly increased overall laying performance compared with birds reared on feed restriction up to maturity.

The results indicate that the improvement in laying performance of birds severely restricted to 9 weeks, but then fed *ad libitum*, could be a consequence of either the delay in sexual maturity (compared with full feeding throughout) improving synchronization of physical and sexual development, or some other permanent effect of early feed restriction upon the endocrine system affecting subsequent rate of lay. The effect of early restriction upon rate of lay is confounded by delayed sexual maturity which means that it is not possible to measure their separate effects on laying performance.

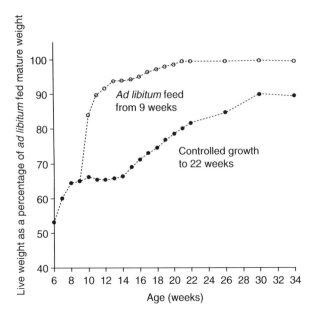

Fig. 6.19. Growth to 34 weeks of Pekin genotype B females reared to achieve 3.07 and 4 kg (75% and 95% of *ad libitum* mature live weight) at 18 weeks of age. All birds were reared on the same feed restriction programme to 9 weeks of age when, they were provided with either *ad libitum* feed or continued feed restriction to 22 weeks of age. All birds were given a constant daylength of 17 h during rearing. (Data from Cherry, 1993.)

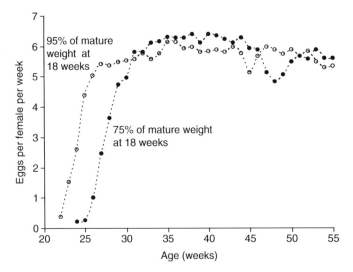

Fig. 6.20. Laying performance of Pekin genotype B females reared to achieve 75% and 95% of *ad libitum* mature live weight at 18 weeks. Birds reared on feed restriction to achieve 75% of mature weight were provided with timed feed restriction of 7 h from 22 weeks; the other rearing treatment continued on *ad libitum* feed. Light programme as in Fig. 6.19. (Data from Cherry, 1993.)

Effects of Age and Live Weight upon Egg Weight

Figure 6.21 describes the relationship between age and egg weight in two trials: genotype A reared in trial 7 to achieve 50%, 65% and 80% of mature weight and genotype B in trial 14 reared to achieve *ad libitum* mature live weight and 75% of mature weight at 22 weeks. Analysis of the separate results provides the following estimates of the effect of age (*x*, weeks) on egg weight between the ages of 30 and 37 weeks in trial 7, and between 25 and 30 weeks in trial 14:

Trial 7 egg weight (g) = 42.1 + 1.30x

Trial 14 egg weight (g) = 22.3 + 2.25x

Egg weight increased steadily with age in both trials but live weight had no substantial effect on egg weight in either trial (see Fig. 6.22). The probable reason for the difference in the relative rate of increase in egg weight between the two trials is that birds in trial 7 were maintained on quantitative feed restriction to 34 weeks before they were allowed 2 h feeding time per day, whereas birds in trial 14 were allowed either *ad libitum* feed or 7 h feeding time from 22 weeks of age.

In another trial, genotype A was grown and then maintained in lay at either 65% or 75% of *ad libitum* mature live weight by quantitative feed restriction. Figure 6.23 shows that egg weight increased with age at about the same rate for both treatments, but maintaining birds at the lower weight significantly reduced weekly average egg weight by 7 g over the period 29–38 weeks of age.

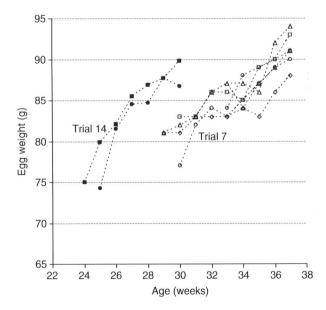

Fig. 6.21. Relationship between age and egg weight for birds reared on alternative growth curves and achieving age at sexual maturity at different ages in trials 7 and 14. Treatments for trial 7 as in Figs 6.10 and 6.13 and treatments for trial 14 as in Figs 6.19 and 6.20. (Data from Cherry, 1993.)

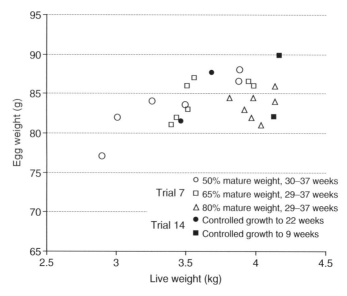

Fig. 6.22. The relationship between live weight and egg weight for birds reared on alternative growth curves in trials 7 and 14. Genotype, controlled growth and lighting programme trial 7, as in Figs 6.10 and 6.13, and trial 14, as in Figs 6.19 and 6.20. In trial 7, live weight and egg weight were recorded weekly from 29 to 37 weeks of age and is the average of birds increased in daylength at either 18 or 24 weeks of age. In trial 14, live weight was recorded at 26 and 30 weeks of age. (Data from Cherry, 1993.)

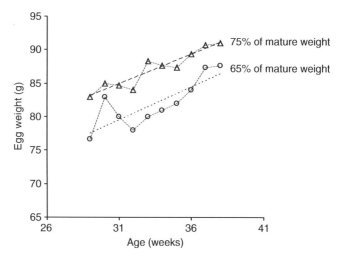

Fig. 6.23. Relationship between age and egg weight for genotype A grown and maintained in lay at 65% and 75% of *ad libitum* live weight by quantitative feed restriction. (Data from Cherry, 1993.)

Analysis of the results provides the following estimates of the effect of age (x, weeks) on egg weight between 29 and 38 weeks of age:

Egg weight (g) 65% of mature weight = 49.0 + 0.934x

Egg weight (g) 75% of mature weight = 57.8 + 0.875x

It was necessary to restrict feed intake to maintain birds at 65% of *ad libitum* mature live weight in lay, and Fig. 6.24 shows that there was a substantial difference in feed intake between treatments. Average daily feed consumption over the period was 210 and 178 g, respectively, for birds maintained at 75% and 65% of mature weight.

In both trials, controlling live weight of birds coming into lay by restricting feed to <210 g per day delayed sexual maturity and limited the increase in egg weight up to 35 weeks of age, compared to birds provided with restricted time feeding from 22 weeks but with feed intake >210 g per day. It is feed income above maintenance which determines egg size, rather than body weight itself.

The effect of egg weight on hatchability, day-old weight and subsequent live weight of table duck will be reviewed in later chapters. Information will be presented to show that it is necessary to restrict energy intake of mature breeding females to limit egg size to less than about 92 g to achieve optimum hatchability. Timed feeding, where birds are allowed limited access to feed, provides a suitable method of limiting feed intake, achieving optimum rate of lay and controlling

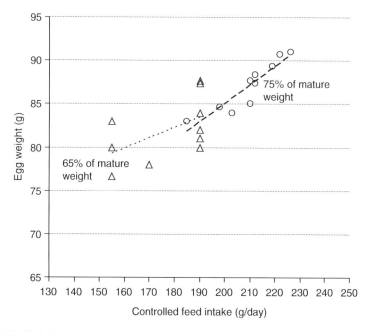

Fig. 6.24. Relationship between feed intake and egg weight for genotype A between 29 and 38 weeks of age. Birds were grown to, and maintained in, lay at 65% or 75% of *ad libitum* mature live weight by quantitative feed restriction. (Data from Cherry, 1993.)

egg weight, but it is essential to use this method of feeding from an early age to enable birds to become familiar with the daily routine. Where birds are reared on either a step-down or a constant long day, timed feeding can be introduced after about 18 weeks of age by providing 2 h feeding time per day, increasing by 1 h a week up to about 7 h in temperate climates.

Providing timed feeding from about 18 weeks allows birds to increase feed intake, gain live weight to about 80% of mature weight and achieve sexual maturity. Increasing daily feeding time gradually by 1 h each week allows birds to increase nutrient intake to satisfy the demands of increasing rate of lay and egg weight. As birds get older and egg weight increases to a level where it can adversely affect hatchability, reducing feeding time provides a method of limiting egg weight to 92 g without adversely affecting rate of lay.

Overview of Restricted Live Weight, Age at Sexual Maturity and Laying Performance

Figure 6.25 summarizes the relationship between live weight at 18 weeks and age at sexual maturity for genotypes A and B reared on a step-down lighting programme. Analysis of results of these trials provides the following estimate of the effect of restricted live weight at 18 weeks, measured as a percentage of *ad libitum* mature live weight (x), upon age at sexual maturity:

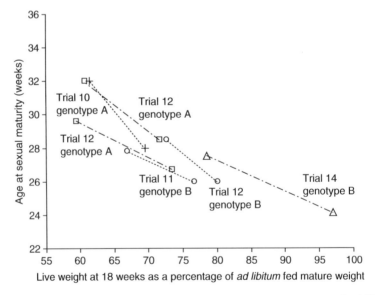

Fig. 6.25. Effect of live weight at 18 weeks expressed as a percentage of *ad libitum* mature live weight on age at sexual maturity, Pekin genotypes A and B with mature weights of 4.94 and 4.1 kg in four trials were given similar step-down lighting programmes during rearing and a 17 h daylength in lay; birds were changed from quantitative feed restriction to timed feeding at different ages. (Data from Cherry, 1993.)

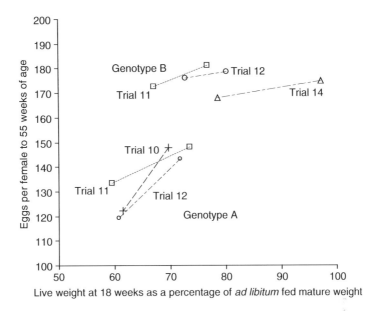

Fig. 6.26. Effect on laying performance in four trials of live weight at 18 weeks expressed as a percentage of *ad libitum* mature weight. Genotypes and lighting programmes as in Fig. 6.25. (Data from Cherry, 1993.)

Age at sexual maturity (weeks) = 42.43 − 0.199x

All trials with both genotypes show a consistent effect of live weight upon age at sexual maturity, with each 10% reduction in live weight associated with 14 days delay in sexual maturity. There is a substantial difference between this estimate and that calculated for the Olver trials reported earlier where restriction did not start until either 3 or 8 weeks of age. Restricting growth is effective in delaying sexual maturity, but the extent to which it can be delayed is limited by the fact that growth of <60% of *ad libitum* mature live weight prevents normal development of the ovary and subsequent sexual maturity.

Increasing live weight in the range 60–95% of *ad libitum* mature live weight at 18 weeks for birds reared on severe restriction from day-old is accompanied by an improved laying performance to a fixed age (see Fig. 6.26) because reducing age at sexual maturity increased the effective laying period (see Fig. 6.20). The difference in the responses of genotypes A and B is probably due to differences in the range over which live weight was measured, affecting age at sexual maturity and subsequent effective laying period to a fixed age of 55 weeks.

Summary of Effects of Feed Restriction on Growth, Age at Sexual Maturity and Laying Performance

Restricted growth delays physiological development, measured in terms of bone and feather growth, behaviour and sexual maturity, probably in relation to the degree of restriction measured as a percentage of *ad libitum* live weight. The potential for further growth remains until development of the ovary has commenced, at which point sexual development takes priority and the capacity for further growth appears to be lost.

Post-mortem examination of birds at sexual maturity and at the conclusion of lay confirms that internal organization of ova is affected by growth to sexual maturity. Feed restriction delays sexual maturity and improves synchronization between growth and sexual development to an extent that significantly reduces, but does not prevent, internal ovulation. This is in contrast to birds reared on *ad libitum* feed, where internal ovulation is frequent and is probably responsible for their inferior laying performance compared to birds reared on restriction.

Several trials have demonstrated that, with the diets used, growth to 56 days and also to 20 weeks was a linear function of restricted energy intake. While it is possible to grow females to less than 50% of *ad libitum* mature live weight, it is necessary to reach 65% in order to achieve sexual maturity, and several trials have shown that birds with time-restricted access to feed seek to achieve and maintain 80% of *ad libitum* mature live weight.

Growth potential can be affected by age, and this may be a consequence of the pattern of the growth of wild and domestic ducks which can achieve mature weight by about 12 weeks of age if food supply is not limiting. Beyond a certain age, which appears to be around 17 weeks in the Pekin duck maintained on long days, and given that live weight >80% of *ad libitum* mature live weight has been achieved, ducks are not able to achieve further growth. Restriction carried out to this age results in a permanent reduction in mature body size. However, where birds are reared on lighting programmes that delay sexual maturity, catch-up growth can occur at a later age.

Temperature affects food intake, growth and subsequent *ad libitum* mature live weight, but does not appear to inhibit physiological development to the same extent as quantitative feed restriction causing the same reduction in live weight. In consequence, in order to successfully delay sexual maturity in hot climates it is probably necessary to relate restriction to *ad libitum* growth in that climate.

Increasing live weight in the range 50–90% of *ad libitum* live weight at 18 weeks, for birds reared on severe feed restriction during the first 8–9 weeks, is accompanied by improved laying performance to a fixed age. This is a consequence of earlier age at sexual maturity extending the effective laying period.

Egg weight is mainly a function of age, rather than body weight, but where quantitative feed restriction is continued into the laying period, limiting feed supply can also reduce egg weight.

Body Composition and Age at Sexual Maturity

The effects of genotype, age, nutrition and other factors on body composition of Pekin duck were described in Chapters 5 (this volume), but only in relation to birds fed *ad libitum* and killed before 56 days.

Figures 6.27 and 6.28 describe the treatments, feed intake and growth recorded in a trial designed to investigate opportunities for altering body composition of parent stock reared on restriction to achieve 65% or 75% of *ad libitum* live weight at 25 weeks of age. Dietary treatments began at day-old and continued until the start of egg production at about 25 weeks. Growth was controlled by adjusting the level of feed restriction on a weekly basis (see Fig. 6.27). Analysis of feed intake (see Table 6.5) shows that, inspite of very wide differences in protein intake growth to 20 weeks of age was principally controlled by energy consumption (x, MJ ME, 0–20 weeks):

20-week live weight (kg) = $1.276 + 0.0095x$

Samples were taken for carcass dissection from all treatments when the birds were 21 and 25 weeks of age. The effects of the treatments upon body composition expressed as a percentage of plucked weight show that the percentage of skin and fat increased with age (Fig. 6.29) while the percentage of breast and leg meat with bone decreased over the same range of plucked body weight (Fig. 6.30). Analysis of carcass dissection data along with that for *ad libitum* fed birds of the same genotype aged 20 weeks provides the following estimates of body composition as a function of plucked body weight (x, kg):

Skin and fat (%) = $-16.5 + 17.6x - 1.133x^2$

Breast and leg muscle with bone (%) = $40.68 - 7.63x + 0.731x^2$

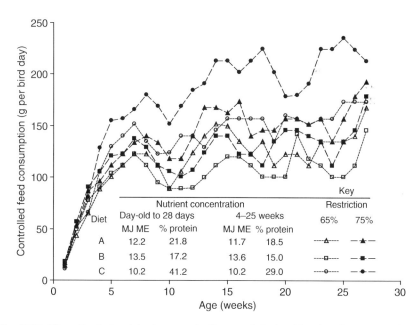

Fig. 6.27. The effect of nutrient concentration on daily feed intake of genotype A reared on feed restriction to achieve 65% and 75% of *ad libitum* mature live weight at 26 weeks. Birds were reared on a step-down, step-up lighting programme with increased daylength from 16 or 20 weeks. (Data from Cherry, 1993.)

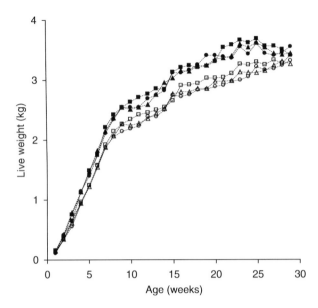

Fig. 6.28. Growth of genotype A, reared on three alternative diets to achieve 65% and 75% of *ad libitum* mature live weight at 26 weeks of age. Treatments as in Fig. 6.27. (Data from Cherry, 1993.)

These results show that the principal factor affecting body composition of females approaching sexual maturity is body weight, and there is little if any opportunity to alter body composition at any specific live weight through manipulation of diet composition. It is therefore not surprising that the dietary treatments had no significant effect upon age at sexual maturity for birds increased in daylength at 16 or 20 weeks when reared to 65% and 75% of mature weight at 26 weeks (see Table 6.6).

Figures 6.31 and 6.32 show the relationship between plucked weight and weight of skin and fat, and breast and leg muscle with bone, expressed

Table 6.5. Energy intake (MJ ME) and controlled growth to 20 weeks of age. (Data from Cherry, 1993.)

Diet	A Starter 12.2 Grower 11.7 (MJ ME/kg)		B Starter 13.5 Grower 13.6 (MJ ME/kg)		C Starter 10.2 Grower 10.2 (MJ ME/kg)	
Target live weight at 26 weeks (% of *ad libitum*)	65	75	65	75	65	75
Energy consumption (MJ ME) to 20 weeks	179	208	182	214	180	232
Live weight at 20 weeks (kg)	2.98	3.35	3.03	3.31	2.89	3.41
Live weight at 20 weeks as a percentage of mature weight	60	68	61	67	59	69

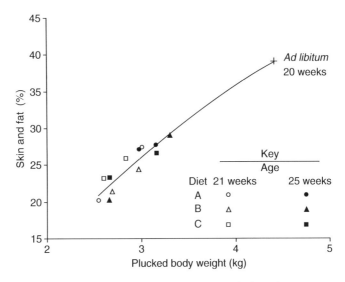

Fig. 6.29. Relationship between plucked body weight (kg) and percentage of skin and fat measured at 21 and 25 weeks of age. Treatments as in Fig. 6.27, and growth to 26 weeks as in Fig. 6.28. (Data from Cherry, 1993.)

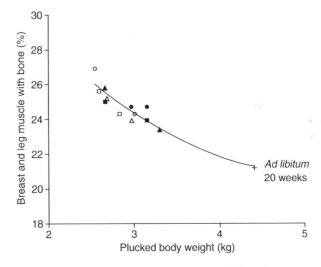

Fig. 6.30. Relationship between plucked body weight (kg) and percentage of breast and leg muscle with bone, measured at 21 and 25 weeks. Treatments as in Figs 6.27 and 6.29. (Data from Cherry, 1993.)

as a percentage for three trials. Birds in these trials were of the same genotype, reared on *ad libitum* and restricted feed using different nutrient specifications and killed at different ages. Analysis of data from these trials provides wider-ranging estimates of the effect of plucked body weight (x, kg) on body composition:

Table 6.6. Live weight and age at sexual maturity for Pekin genotype A reared on alternative diets and controlled growth to 26 weeks of age with daylength increased at 16 or 20 weeks. (Data from Cherry, 1993.)

	Diet A		Diet B		Diet C	
Increase in daylength at 16 weeks						
Live weight at 20 weeks (kg)	2.96	3.33	2.97	3.28	2.99	3.42
Live weight at 26 weeks (kg)	3.2	3.54	3.25	3.57	3.18	3.35
Percentage of mature live weight	65	72	66	72	64	68
Age at sexual maturity (weeks)	29.3	28.5	29.0	28.3	29.0	28.0
Increase in daylength at 20 weeks						
Live weight at 20 weeks (kg)	2.99	3.37	3.09	3.34	2.79	3.4
Live weight at 26 weeks (kg)	3.12	3.55	3.27	3.57	3.16	3.53
Percentage of mature live weight	63	72	66	72	64	71
Age at sexual maturity (weeks)	30.5	29.8	30.0	29.5	30.5	29.8

Skin and fat (%) $= -11.28 + 14.57x - 0.689x^2$

Breast and leg muscle with bone (%) $= 42.1 - 8.67x + 0.895x^2$

With the exception of body protein of birds grown to achieve 50% of *ad libitum* live weight at 59 days, where protein intake was probably insufficient to support bone and muscle growth (this point is not included in the above estimate), the data clearly show that neither age nor *ad libitum* or restricted feed or alternative nutrient specifications during rearing had any noticeable effect on body composition at a given weight. The proportions of the Pekin duck appear to be determined by growth, with body composition being a simple function of size alone.

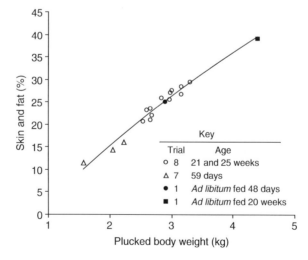

Fig. 6.31. Relationship between plucked body weight and percentage of skin and fat for genotype A recorded at different ages in three separate trials. (Data from Cherry, 1993.)

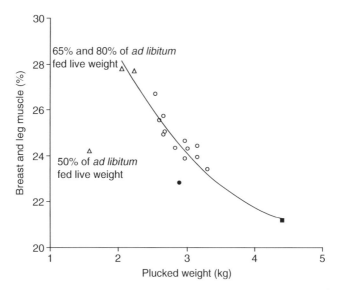

Fig. 6.32. Relationship between plucked body weight (kg) and percentage of breast muscle and leg muscle with bone for genotype A recorded in three trials. Key as in Fig. 6.31. (Data from Cherry, 1993.)

A possible explanation for this relationship is that despite domestication over several thousand years the domestic Pekin still shares many similarities with its wild ancestor the Mallard. Body composition is of critical importance in wildfowl and subject to continuous selection pressure because the relationship between breast muscle, body weight and wing area affect flight. As described in Chapter 5 (this volume), while there is a substantial difference in absolute growth rate between wild Mallard and domestic Pekin, the time and pattern of growth to somatic maturity are similar. Significantly, the relationships between body weight and skin and fat, and breast and leg muscle when measured as a percentage of mature weight are also very similar (Gille and Salomon, 1998).

Fledging in dabbling ducks occurs at about 52 days, but critically depends on the integration of growth and calcification of the sternum, to provide a base for the attachment and growth of breast muscle, and also on the growth and maturation of both primary and secondary flight feathers. Domestication has significantly reduced the absolute size of primary flight feathers but the time and pattern of growth of primary wing feather, when expressed as a percentage of the mature length, are very similar to those in Mallard.

Wild Mallard in common with other migratory wildfowl seek to increase body fat to meet the energy requirement for migration and subsequent reproduction in habitats where energy supplies are scarce at the time of nesting. Krapu (1981) in an extensive study reported that reserves of body fat contribute substantially towards reproduction; he also reported that weight loss averages 25% from pre-laying to late incubation for wild Mallard in the pothole region of North Delaware. This has been confirmed by Pattenden and Boag (1989) for game farm Mallard maintained on feed restriction; they also reported that body weight of both sexes was positively correlated with body fat. Juvenile Mallard seek to store fat prior to sexual maturity and the constraint controlling this function is the relationship between live weight and the ability to fly. This has been described as 'fitness' and measured in terms of wing load (Poole, 1938) and wing area/weight ratio (Prince, 1979). Analysis of the data recorded by Poole (1938) of body weight and wing area of common species of American wildfowl shows a good relationship, with wing area (cm^2) = W(g)$^{0.96}$.

Summary of the Effects of Body Composition on Age at Sexual Maturity

Consideration of the effect of body composition on sexual maturity and laying performance is to some extent confounded by its relationship to body weight. The evidence from Fig. 6.13 shows that females grown and maintained at 50% of *ad libitum* live weight did not achieve sexual maturity, which could have been a result of either failing to achieve a minimum threshold of fatness measured as a percentage of live weight or, alternatively, there may be a minimum lean mass essential to achieve sexual maturity. Figures 6.13 and 6.15 indicate that the minimum live weight necessary to achieve sexual maturity appears to be at least 65% of *ad libitum* mature live weight and 75% is essential to maintain maximum rate of lay. The body composition achieved at these weights might also be considered essential minimum values for Pekin females to achieve sexual maturity and maintain optimum rate of lay.

There is a very widely held view that increasing live weight and percentage of body fat is highly likely to substantially reduce laying performance of parent stock. However, Fig. 6.20 shows that where ducks were reared on feed restriction to 9 weeks, and then subsequently allowed, by provision of *ad libitum* feed, to achieve 95% *ad libitum* live weight at 18 weeks, although almost certainly very much fatter than fully restricted birds at sexual maturity, they were able to achieve similar rates of lay. This indicates that the increased rate of lay obtained from birds reared on restriction compared to birds reared on *ad libitum* feed from day-old is not a consequence of differences in percentage of body fat at sexual maturity but is probably a result of delayed sexual maturity.

It appears that the domestic duck is able to achieve sexual maturity and maintain a satisfactory rate of lay over a range of body compositions, and opportunities for altering body composition by feed restriction are limited by the fixed relationship of body composition to body weight.

Effect of Natural Daylight upon Age at Sexual Maturity

In view of the space requirements of growing parent stock (0.45 m^2 per bird to sexual maturity) and the effect this has on the cost of providing lightproof rearing accommodation, it is normal practice in many parts of the world to rear ducks extensively, or in low-cost windowed accommodation, so that birds are grown on natural daylength, which varies according to latitude and time of year.

Figure 6.33 describes the change in daylength during rearing between 5 and 26 weeks for ten commercial genotype B parent flocks located at 52°N in the UK, hatched at different times of the year and given artificial increases in daylength from 18 weeks to achieve 17 h daylength at 26 weeks of age.

Figure 6.34 shows the effect of the month that parent stock were hatched on subsequent age at sexual maturity and laying performance to a fixed age of 50 weeks. Figure 6.35 describes the effect of hatch date and change in daylength between 5 and 26 weeks (x, h) on age at sexual maturity, and analysis of these data provides the following estimates:

Age at sexual maturity (weeks) $= 28.39 - 0.256x + 0.004x^2$

Analysis of the effect of age at sexual maturity (y, weeks) on subsequent laying performance to a fixed age of 50 weeks gives:

Eggs per female to 50 weeks $= 314.9 - 6.46y$

Figure 6.33 shows that flocks hatched in the period from March to May were exposed to only small changes in daylength to 26 weeks of age and consequently had similar ages at sexual maturity (*see* Fig. 6.34). Flocks hatched from June to September received a step-down in natural daylength to 18 weeks,

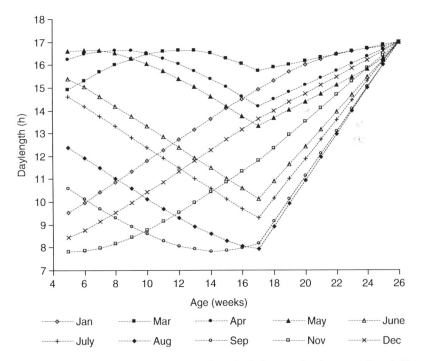

Fig. 6.33. Effect of month of hatch on daylength during rearing between 5 and 26 weeks for flocks reared at 52°N with 23 h light from day-old to 5 weeks, followed by natural days to 18 weeks, after which daylength was increased in equal steps using artificial light to provide 17 h daylength at 26 weeks of age.

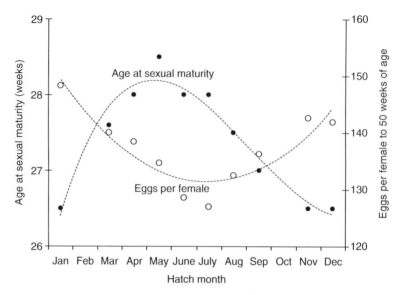

Fig. 6.34. Relationship between month of hatch, age at sexual maturity and laying performance for commercial parent flocks reared at 52°N with natural daylight from 5 to 17 weeks. Lighting programmes as described in Fig. 6.33. Birds of genotype B were reared on feed restriction to achieve about 75% of *ad libitum* mature live weight at 26 weeks. (Data from Cherry, 1993.)

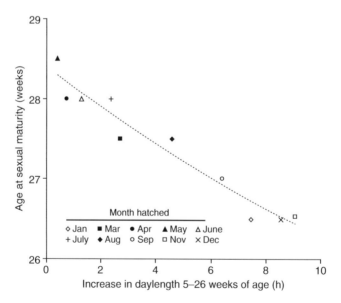

Fig. 6.35. Effect of month of hatch on increase in daylength between 5 and 26 weeks and subsequent age at sexual maturity of commercial flocks of parent stock. Genotype, lighting programme and controlled growth as in Figs 6.33 and 6.34. (Data from Cherry, 1993.)

followed by a step-up in daylength of 7–9 h between 18 and 26 weeks. Flocks hatched in November, December and January received a gradual increase in daylength from 5 weeks onwards.

It appears that although birds hatched from June to August received a step-up of 7–9 h between 18 and 26 weeks, this stimulus was largely cancelled out by the large step-down in natural daylength from 5 to 18 weeks and, in consequence, they had a similar age at sexual maturity to birds hatched from March to May, which received only a small step-up in daylength from 18 weeks. Both groups were later maturing than flocks hatched from September to January, which received an overall increase in daylength of 6–9 h between 5 and 26 weeks of age.

Flocks hatched from June to December, which received a large increase in daylength between 18 and 26 weeks, showed a reduction in rate of lay following peak production (see Fig. 6.36). This is in contrast to birds hatched from January to May, which received a small decrease in daylength between 5 and 18 weeks and small increases thereafter.

The December- and May-hatched flocks both received an increase in daylength of about 3 h between 18 and 26 weeks, but only the December flock showed a post-peak reduction in lay, probably because this flock had an increase of about 5 h in natural daylength from 5 to 18 weeks of age, in contrast to the May flock which received a step-down of about 3 h to 18 weeks of age.

Empirical evidence gained with Pekin parent stock in Europe and with both Khaki Campbell parent stock and hybrid layers supplied to the Far East shows that parent stock provided with a single step-up of several hours in daylength at about 22 weeks experience a similar, or greater, temporary reduction in post-peak rate of lay. A possible explanation for the effect of step-up lighting on post-peak laying performance is provided later in this chapter.

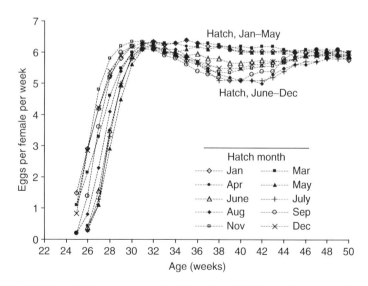

Fig. 6.36. Relationship between month of hatch and laying performance of commercial breeding flocks. Genotype, feed restriction and lighting programme as in Figs 6.33 and 6.34. (Data from Cherry, 1993.)

Step-up Lighting Programmes

Step-up lighting programmes can vary in both the age at which step-up lighting is introduced and the rate of increase in daylength. The programme can be as simple as that sometimes used in Eastern Europe, where parent stock are given a single step-up from natural daylength to 16h. For parent stock hatched in July and August this can mean a single jump of 8–9h when placed into breeder accommodation in January at about 24 weeks of age.

In the USA, European Community (EC) and Eastern Europe, step-up light programmes vary a great deal, with step-up lighting commencing as early as 16 weeks. Both equal and variable increments in artificial light are used to achieve 16–17h daylength between 21 and 28 weeks of age.

Figure 6.37 shows the significant effect of age at increase in daylength on age at sexual maturity in trials which used step-down, step-up light programmes. Birds in these trials were grown on different growth curves and allowed access to timed feed restriction at different ages, which is responsible for some of the differences in age at sexual maturity between trials.

Analysis of the results provides the following estimates of the effects of age at increase in daylength (x, weeks) and live weight at 20 weeks (y, measured as a percentage of *ad libitum* mature live weight) upon age at sexual maturity:

$$\text{Age at sexual maturity in weeks} = 26.6 + 0.419x - 0.08y$$

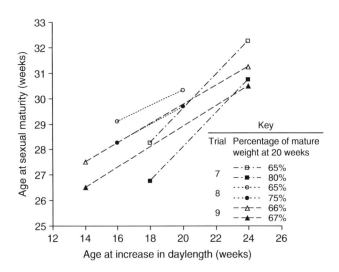

Fig. 6.37. Effect of age at increase in daylength and of body weight at 20 weeks on age at sexual maturity for birds reared on a step-down, step-up light programme. Genotype A in three trials were given 23h daylength to either 7 or 8 weeks when light was reduced to 8h. At 14, 18, 20 or 24 weeks of age, daylength was increased by 1h each week up to 17h. All birds were reared with restricted feed until 26 weeks of age. (Data from Cherry, 1993.)

Step-down, step-up Lighting

Parent stock are frequently given 23h daylength from day-old for up to about 8 weeks, based on the widespread belief that this reduces early mortality. Figure 6.38 shows that cutting daylength from 23 to 8h at 8 weeks delays maturity by about a week, compared to birds given 8h from day-old. However, the step-down at 8 weeks had no significant effect upon laying performance to 55 weeks of age. This trial also showed that increasing daylength by 1h per week from 14 weeks reduces age at sexual maturity by 4 weeks, compared to birds kept on short days until 24 weeks, and increased laying performance to 55 weeks of age by about 14 eggs per female.

Using a short daylength during rearing and delaying the first increase of daylength for birds reared on a step-up light programme provides an opportunity to delay sexual maturity by several weeks, and so cuts out the production of many eggs too small to incubate. However, this requires access to expensive lightproof rearing accommodation to maintain birds on the short daylength prior to introducing a step-up light programme. Alternatives which allow birds to be reared in windowed accommodation and yet avoid increases in daylength, which would be likely to reduce age at sexual maturity, are either to rear birds on a step-down light programme or to provide a constant daylength such as 17h from day-old to the end of lay.

Figure 6.39 describes the effect of rearing genotypes A and B on either a step-up or a step-down rearing light programme.

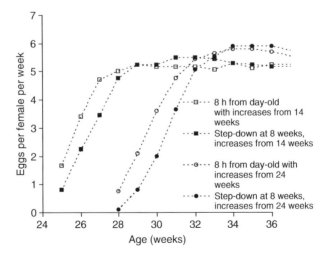

Fig. 6.38. The effect of either a step-down from 23 to 8h light at 8 weeks or a constant 8h light upon age at sexual maturity, for birds given increasing daylength (1h each week up to 17h) from either 14 or 24 weeks. Genotype A was fed to achieve about 68% of mature weight at 20 weeks of age. (Data from Cherry, 1993.)

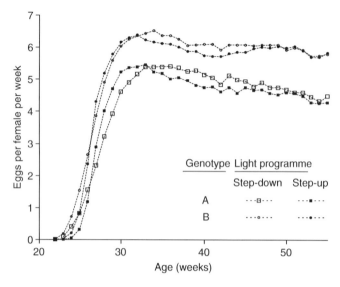

Fig. 6.39. Effect of rearing birds on two light programmes on sexual maturity and laying performance. Genotypes A and B were reared on either a step-down, step-up light programme of 23 h to 8 weeks and 8 h from 8 to 18 weeks, when daylength was increased by 1 h a week to 17 h, or on a step-down programme of 23 h to 8 weeks when daylength was reduced in equal steps to 17 h at 18 weeks of age. Birds were reared to achieve 75% of *ad libitum* mature live weight at either 20 or 26 weeks of age. (Data from Cherry, 1993.)

There was no significant difference between the two rearing light treatments in age at sexual maturity for birds grown to achieve about 75% of *ad libitum* mature live weight at 20 weeks of age (see Fig. 6.40), but the step-down light significantly increased age at sexual maturity for birds reared to achieve 75% of mature weight at the later age of 26 weeks.

The results of this and another trial (see Fig. 6.41) show that live weight can interact with rearing light programme to affect age at sexual maturity. Live weight in the range 60–65% of *ad libitum* mature live weight at 20 weeks had significantly greater effect upon age at sexual maturity for ducks grown on a step-down lighting programme than for birds reared on a step-up lighting programme. Birds in trial 11 were given access to timed feed from 26 weeks of age, but birds in trial 10 were maintained on quantitative feed restriction until 29 weeks; this delayed sexual maturity of all treatments in the trial by about 2 weeks. However, Fig. 6.41 shows that the interaction between live weight and rearing light programmes was similar in both trials.

The difference between step-up and step-down lighting treatments on age at sexual maturity is increased by approximately 0.8 days for each reduction of 1% in 20-week live weight in the range 59–74%, measured as a percentage of *ad libitum* mature live weight, but this may not be a linear effect.

Controlling growth to maturity represents the principal way in which sexual maturity can be delayed and laying performance improved for Pekin ducks

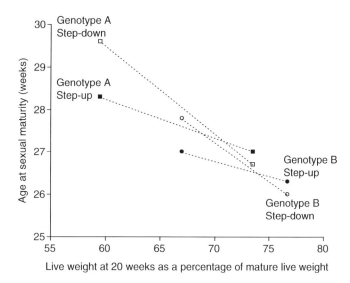

Fig. 6.40. Effect of rearing light programme and live weight expressed as a percentage of *ad libitum* mature live weight upon age at sexual maturity. Genotype, controlled growth and light programme as in Fig. 6.39. (Data from Cherry, 1993.)

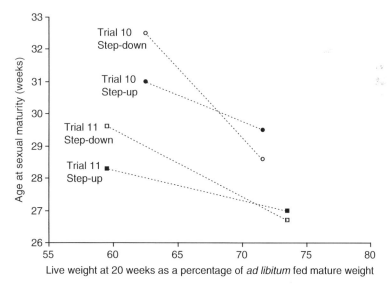

Fig. 6.41. Effect of live weight at 20 weeks and rearing light programme on age at sexual maturity. Genotype A in two trials were reared on either a step-down, step-up light programme or on a step-down light programme. Birds in both trials were reared on similar levels of feed restriction to 20 weeks, but birds in trial 10 were maintained on quantitative feed restriction to 29 weeks. (Data from Cherry, 1993.)

reared on a non-stimulatory rearing light programme, in contrast to ducks reared on a step-up light programme, where age at increase in daylength can also be used to delay sexual maturity. However, rigorously controlling nutrient intake when birds are coming into lay, using either quantitative feed restriction or time-controlled access to feed, can also delay age at sexual maturity of birds grown on either step-up or step-down rearing light programmes. This represents a further method by which sexual maturity can be controlled, but empirical evidence suggests that delaying sexual maturity in this manner can reduce subsequent rate of lay and affect both fertility and hatchability.

In these trials rearing light programme had no significant effect on overall laying performance to 55 weeks for either genotype, but had a small effect upon rate of lay following peak production. Both genotypes reared on the step-up light programme showed a lower rate of lay from 35 to 45 weeks than birds reared on the step-down light programme (see Fig. 6.39).

Lighting treatments had no significant effect on egg weight or hatchability of fertile eggs, but the step-up rearing light programme improved early fertility to about 28 weeks for both genotypes by a small but significant amount. The effects of controlled growth, rearing light programme and controlled feed when birds are coming into lay on fertility and hatchability will be further reviewed in Chapter 8 (this volume).

There was no significant difference in age at sexual maturity or overall breeding performance in trials or in commercial flocks of genotype B when reared on either a step-down to 17 h or a constant 17 h daylength. This indicates that the duck is not sensitive to changes in daylength between 24 and 17 h.

Constant Lighting

Figure 6.42 describes the laying performance of six large commercial trial flocks where genotype B ducks were hatched at intervals between March and November, reared to achieve about 75% of *ad libitum* mature live weight at about 20 weeks of age and given a constant daylength of 17 h from day-old to end of lay. Age at first egg was about 21 weeks and birds reached 50% lay by about 27 weeks. All flocks, regardless of hatch date, maintained a similar and satisfactory rate of lay with no temporary reduction in post-peak rate of lay.

In view of the simplicity and effectiveness of this constant light programme, and its suitability for all latitudes and all seasons, it deserves to be adopted for all meat strain breeding flocks reared with controlled feeding.

Effect of Alternative Lighting Programmes During Rearing on Male Fertility

The trials reported have described the effects of daylength during rearing on sexual maturity of females. Males in those trials were reared with females and given the same lighting and feed restriction programme.

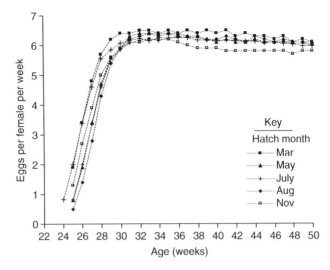

Fig. 6.42. Laying performance of six commercial flocks of genotype B hatched between March and November and given a constant 17h daylength from day-old to the end of lay. Birds were reared on feed restriction to achieve about 75% of *ad libitum* mature live weight at 20 weeks of age. Average laying performance to 50 weeks was 147 eggs per female, based on number alive at 20 weeks of age. (Data from Cherry, 1993.)

Table 6.7 describes the effect on fertility of rearing males on four light programmes and two growth curves. Drakes similar to genotype A were reared on two controlled growth treatments to achieve either 70% or 75% of *ad libitum* mature live weight by 18 weeks of age. Genotype B females were reared to achieve about 80% of *ad libitum* mature live weight at 20 weeks of age and given a constant daylength of 17h from day-old to end of lay. Males were housed with females at 18 weeks, when all treatments were given a 17h day and approximately 200g of feed per bird per day. By 20 weeks males achieved about 78% of *ad libitum* mature live weight regardless of the growing treatment to 18 weeks.

Table 6.7 shows that both growth to 18 weeks and lighting programme had some effect on fertility at 25 weeks but, by 27 weeks, neither growth nor lighting programme had any significant effect upon male fertility. The results of this trial along with empirical evidence from field-scale trials where males were grown to achieve about 80% of mature weight at 20 weeks show that neither step-up nor step-down nor a constant daylength of 17h had any significant effect on male fertility. Factors affecting fertility and hatchability are further reviewed in Chapter 8 (this volume).

Rearing Light Programmes and Post-peak Laying Performance

Analysis of commercial records (see Fig. 6.36) and trial results (see Fig. 6.39) shows that giving a step-up in daylength of more than about 3h when Pekin

Table 6.7. Effect of restricted growth and lighting programme upon male fertility. (Data from Cherry, 1993.)

Lighting programme		Live weight at 18 weeks as a percentage of mature weight	Percentage of infertile eggs Age (weeks)						
Day-old to 6 weeks	6–17 weeks		25	26	27	28	29	30	31
17h	Constant 17h	69	23	10	8	10	8	7	7
		74	25	10	11	10	8	7	7
		Mean	24	10	10	10	8	7	7
23h	Step-down from 23 to 17h	69	41	11	9	8	6	6	6
		74	19	11	10	10	7	8	6
		Mean	30	11	10	9	7	7	6
17h	Step-down to 8h, then step-up to 17h	69	19	9	14	8	8	8	8
		74	18	10	9	8	8	7	8
		Mean	19	10	12	9	8	8	8
17h	Step-down to 12h, then step-up to 17h	69	23	11	13	8	7	8	7
		74	11	9	7	7	7	7	8
		Mean	17	10	10	8	7	7	8

females are approaching sexual maturity appears to temporarily reduce post-peak rate of lay compared with birds reared on either a step-down to 17 h or a constant daylength of 17 h. A possible explanation for this effect can be found in a comprehensive review by Assenmacher (1974) of external and internal components of the mechanism controlling the reproductive cycles of drakes.

Assenmacher (1974) described an experiment by Benoit which showed that temperature did not interfere with light stimulation of the gonads, by conducting an experiment in which Pekin drakes were maintained in a cold environment and provided with an increasing daylength which stimulated them to become sexually active, whereas drakes maintained in a hot environment but on short days did not. This showed that the principal reason that ducks reproduce in springtime is due to an increase in daylength and not because of increased temperature.

In another experiment, Benoit maintained Pekin drakes on natural daylength (latitude 49°N) and X-ray examination of the testes at the beginning of each month over a full year showed that growth of the testes commenced shortly after the winter solstice to reach a maximum dimension by April. Testicular regression along with reduced plasma testosterone commenced in May before the summer solstice and was complete by August. The results of this experiment have since been confirmed for Pekin ducks reared in the southern hemisphere by Cardinali *et al.* (1971) but with a 6-month phase shift corresponding to seasonal difference.

Sharpe (1984a) describes this response as 'photorefractory', being absolute when birds show regression of sex organs despite continued increase in daylength, which is the case for Mallard and the majority of both dabbling and diving ducks. However, Assenmacher (1974) describes how Benoit demonstrated with Pekins that maintaining daylength in June by artificial light could induce testicular growth back to maximal size, which has been described by Sharpe (1984b) as 'relative photorefractoriness'.

The results of these experiments explain why commercial genotype B flocks reared on natural daylength from 5 to 18 weeks, which were exposed to an increase in daylength of more than 3 h between 18 and 26 weeks, and genotype A reared on 8 h to between 14 and 24 weeks and then increased to 17 h in weekly steps all showed a decline in post-peak rate of lay, compared to birds reared on either a step-down or a constant daylength of 17 h. Ducks when stimulated by a step-up light programme (either natural or artificial) at an age when they are particularly photosensitive become 'relatively photorefractory' as a result of the large increase in daylength. Similarly it may also explain the high level of broodiness observed in June–December hatches and in several trials where birds were given step-up light programmes.

Summary of Effects of Daylength upon Age at Sexual Maturity

Changes in natural or artificial daylength during rearing can significantly affect age at sexual maturity. Increasing daylength during rearing or reducing the age at which artificial light is first increased will reduce age at sexual maturity in a linear manner, if the ducks are given enough feed to allow ovarian development.

Rearing Pekin parent stock on a light programme stepping down from 23 to 17 h at 18 weeks when compared to a constant daylength of 17 h from day-old produces a similar age at sexual maturity for birds of the same live weight. This shows that Pekin ducks are not affected by changes in daylength in the range 17–24 h.

There can be a significant interaction between rearing light programme and feeding regime, as reflected in live weight at 20 weeks. If feed restriction is severe, light treatments have a greater effect upon reproductive development.

Rearing light programme can affect not only age at sexual maturity but also subsequent rate of lay. Increases in natural or artificial daylength of more than 3 h during the latter stages of rearing can reduce post-peak laying performance. This is described as a 'photorefractory response'.

Excellent results have been obtained by maintaining meat strain ducks and drakes on a constant photoperiod of 17 h from day-old until the end of the breeding cycle. Since this programme is simple and applicable at all latitudes it deserves to be adopted as the standard method for rearing Pekin breeding stock.

Rearing light programme can significantly affect male sexual maturity, but there appears to be no advantage measured in terms of improved early fertility in rearing males on a different light programme from females.

Temperature, Controlled Growth and Age at Sexual Maturity

A small-scale trial carried out in Singapore (see Fig. 6.11) showed that while high ambient temperature reduced growth rate and mature live weight, it produced a smaller delay in age at sexual maturity (see Table 6.4) than feed restriction at normal temperature applied to give a comparable reduction in live weight. At 20 weeks of age females reared on feed restriction in Singapore were approximately 70% of the weight of birds grown on *ad libitum* feed in the UK at about 11°C, but almost 90% of the Singapore *ad libitum* live weight. They achieved sexual maturity at 24 weeks of age, at least 2 weeks earlier than restricted birds of the same genotype reared on a step-down lighting programme in the UK.

Table 6.8 describes a trial designed to investigate opportunities for increasing age at sexual maturity of birds reared at high ambient temperature, by controlling growth to 18 weeks to achieve 75% and 85% of locally derived *ad libitum* mature live weight. Pekin females similar to genotype B, reared on both *ad libitum* and controlled growth and given a constant daylength of 17 h, achieved sexual maturity at about the same age of 27–28 weeks, but controlled growth dramatically improved rate of lay (see Fig. 6.43). The two groups reared on feed restriction were very similar in age at sexual maturity and laying performance, probably because, although each treatment followed a separate pathway to 18 weeks, as soon as the birds were given *ad libitum* feed at 20 weeks they rapidly gained weight (see Table 6.8).

Birds reared on *ad libitum* feed laid their first egg before 20 weeks, but did not achieve 50% lay until about the same age as restricted birds. This was probably due to internal ovulation.

Table 6.8. Effect of controlled growth on age at sexual maturity and laying performance for birds reared at high ambient temperature (27°C) and given a constant daylength of 17h. (Data from Cherry, 1993.)

Treatment	Weight at 18 weeks (kg (%))	Weight at 22 weeks (kg (%))	Age at sexual maturity (weeks)	Eggs per female to 60 weeks	Feed intake 30–60 weeks (g per day)	Mean egg weight 30–60 weeks (g)
Ad libitum	3.5 (100)	3.5 (100)	28	111	185	85
Controlled growth 85% at 18 weeks	3.0 (86)	3.3 (94)	27	185		
Controlled growth 75% at 18 weeks	2.6 (74)	3.3 (94)	27.5	182		
Average controlled growth treatments					192	91

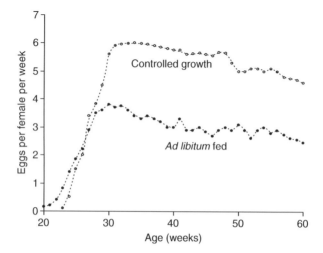

Fig. 6.43. Laying performance of genotype B grown at high ambient temperature (27°C) and fed *ad libitum* or restricted to achieve 75% or 85% of *ad libitum* live weight at 18 weeks of age. Laying performance of the two restricted treatments was very similar and is shown as an average. Birds were given a constant daylength of 17 h, daily feed allowance was increased from 18 weeks and all birds were given *ad libitum* feed from 20 weeks of age.

The results of this trial show that both increased temperature and feed restriction can reduce growth and delay sexual maturity relative to birds grown in a temperate climate, but only delayed sexual maturity by feed restriction improves subsequent rate of lay. The evidence from this trial suggests that to avoid premature age at sexual maturity for birds grown at high ambient temperature and given a constant daylength, it is necessary to relate controlled growth to that of *ad libitum* fed birds reared in similar environmental conditions.

Effect of Age and Weight at Sexual Maturity upon Egg Production

Figure 6.44 summarizes the relationship found, in two genotypes in seven trials investigating the effects of controlled growth, body composition and lighting programme during rearing, between age at sexual maturity and subsequent laying performance. Delaying sexual maturity beyond about 26 weeks reduces laying performance to a fixed finishing age for both genotypes.

Analysis of the various factors which have been shown to influence age at sexual maturity and subsequent laying performance is hampered by the fact that in several trials quantitative feed restriction of some treatments was continued to between 26 and 30 weeks of age, either to control growth or to maintain a specific live weight in lay. The effect of this continued restriction was not only to increase age at sexual maturity of these treatments, but also to reduce early rate of lay and substantially diminish overall laying performance, and this is responsible for some of the variation in the apparent effect

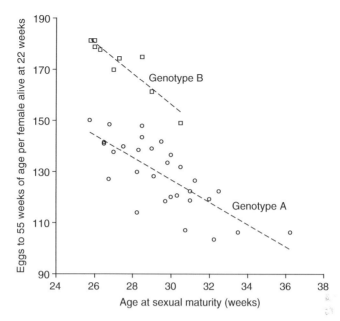

Fig. 6.44. Relationship between age at sexual maturity and laying performance to a fixed age of 55 weeks. Genotypes A and B were reared in seven different trials on various feed restriction and lighting programmes and given different feed allowances when coming into lay. (Data from Cherry, 1993.)

of age at sexual maturity upon laying performance in Fig. 6.44. However, when constants were fitted to remove mean differences between trials arising from these and other effects, live weight at 20 weeks (x, measured as a percentage of *ad libitum* mature live weight) had a significant effect upon laying performance.

Eggs per female to 55 weeks of age $= 62.11 + 0.65x$

Several trials have also shown that breeding ducks seek to achieve a minimum live weight in lay of at least 80% of *ad libitum* mature live weight. Ducks can only be maintained beneath this weight during lay by continued severe feed restriction, which not only delays sexual maturity but also directly reduces rate of lay and overall laying performance.

Growing breeding ducks of these genotypes to reach approximately 80% of *ad libitum* mature live weight at 18–20 weeks is most likely to achieve the optimum benefits of feed restriction, increasing laying performance to a fixed age, and reducing feed cost during rearing.

Analysis of the effect of rearing light programme on subsequent laying performance shows there is little benefit to be gained by using step-up light programmes because trials, commercial performance in the UK and empirical experience within the EC and Eastern Europe show that increasing daylength by more than about 3h during the latter stages of rearing can substantially reduce post-peak rate of lay.

There appears to be no difference between the effect of a step-down light programme (to 17 h) and a constant daylength of 17 h during rearing upon age at sexual maturity and subsequent laying performance. Both trials and empirical experience show that ducks reared on a constant 17 h daylength and grown to achieve 80% of *ad libitum* live weight by 20 weeks can achieve sexual maturity by 26 weeks and produce a consistent and high rate of lay to 55 weeks (see Fig. 6.42). This, along with the economic advantage of using windowed, naturally ventilated housing and natural light, and with the practical advantage of not having to regularly adjust time switches, suggests that the optimum lighting programme for breeding ducks is to provide a constant daylength of 17 h from day-old to the end of lay.

Several trials have shown that it is possible to delay age at sexual maturity by quantitative feed restriction from 20 to 30 weeks of age, but this method of delaying sexual maturity can significantly affect subsequent rate of lay and reduce egg weight (see Figs 6.23 and 6.24). In view of the evidence of the effect of age on growth there appears to be no reason to restrict feed consumption beyond 20 weeks of age (see Figs 6.14 and 6.16).

Both trials and commercial experience show that to achieve optimum rate of lay and maximum early egg weight it is absolutely essential to increase feed allowance when birds are coming into lay. A simple and effective method is to provide 2 h access per day to feed from 18 weeks, increasing by 1 h a week to between 6 and 8 h in temperate climates. In either hot or cold climates it is necessary to substantially increase feeding time, and as ambient temperature approaches and remains near 0°C or above about 24°C it is usually necessary to supply feed *ad libitum*.

Effect of Age at Sexual Maturity on Egg Weight

Age at sexual maturity had no significant effect on egg weight (at a specified age) in any of the trials reported and Fig. 6.21 shows that in two trials where age at sexual maturity varied between 26 and 33 weeks, current age was the significant factor affecting egg weight.

Figure 6.45 illustrates the effect of age on egg weight of both meat and egg-laying genotypes and shows that egg weight increases by about 1.75 g per week between 25 and 32 weeks of age, irrespective of genotype, climate, body weight or method of feeding. This means that early maturing flocks have a lower egg weight to 30 weeks of age, which can affect hatchability, day-old weight and subsequent growth, and needs to be taken into account when considering optimum economic age to achieve sexual maturity. Commercial experience indicates that ducks hatched from eggs smaller than about 70 g are less viable, because they have difficulty in using standard feeding and drinking equipment and in competing in a commercial environment with ducks hatched from larger eggs.

Figure 6.46 describes the effect of age upon egg weight and the percentage of eggs <70 and >105 g and unsuitable for incubation produced by genotype B in a large-scale trial. Females were reared to achieve 80% of *ad libitum*

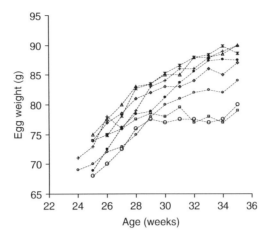

Fig. 6.45. Relationship between age and egg weight for both meat and egg-laying genotypes located in cold, temperate and tropical climates reared on restriction to between 18 and 26 weeks and given either *ad libitum*, quantity or timed feed restriction in lay.

mature live weight at 20 weeks with a constant daylength of 17h, and then given 2h feeding time per day at 20 weeks, increasing by 1h per week to 6h. They reached 50% lay at about 26 weeks of age and at 25 weeks approximately 35% of all eggs produced were too small for incubation. Commercial records of the same genotype reared in the same manner show that at 24 weeks more than 50% of eggs were rejected. This along with the effects of

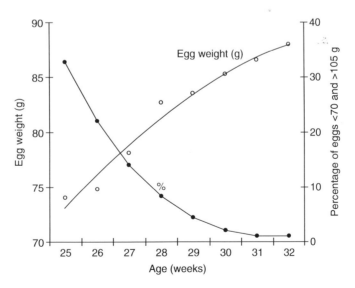

Fig. 6.46. Relationship between age, egg weight and percentage of eggs <70 and >105 g, and thus unsuitable for incubation. (Data from Cherry, 1993.)

parent age on fertility and hatchability (see Chapter 8, this volume) suggests that there is no economic advantage in producing hatching eggs from Pekin parent stock before 24 weeks.

The Effect of Egg Weight on Growth

Egg weight affects day-old weight and subsequent growth and needs to be taken into account when considering optimum economic age to achieve sexual maturity. Figure 6.47 describes the results of a trial designed to investigate the effect of egg weight on day-old weight and subsequent growth. Analysis of the data provides the following estimates, x being the egg weight in grams:

Day-old weight (g) = $0.645x$

43-day weight (kg) = $2.662 + 0.0086x$

Egg weight significantly affected day-old and 43-day weight. Day-old weight increased by 0.64 g for each increase in egg weight of 1 g, which is close to the 0.58 g recorded for ducks by Shanawany (1987).

The analysis shows that 43-day weight increased by 8.6 g for each 1 g increase in egg weight, but this may overestimate the effect of egg weight changes due to age. The estimate is based upon within-flock comparison and, in any 1 week, the largest eggs will tend to come from the larger ducks in the flock. The genetic

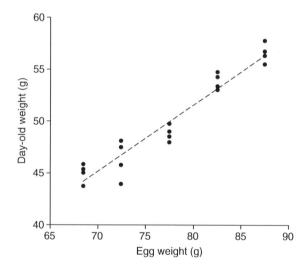

Fig. 6.47. Effect of egg weight on day-old weight. Eggs from two parent flocks of genotypes A and B achieving sexual maturity at 27 and 29 weeks and 26 and 27.5 weeks, respectively, were collected for about 14 days, weighed, identified to supply flock, separated into 5 g classes and placed in the same incubator until hatching. Day-olds from the four flocks and five egg weight divisions were weighed and identified to treatment. At 43 days all birds were weighed individually and recorded to treatment and sex. (Data from Cherry, 1993.)

correlations both between maternal size and egg weight and between maternal size and progeny potential for growth contribute to growth to 43 days.

There is a great deal of experimental evidence of the effect of egg weight upon day-old weight and subsequent growth of broiler chicken (Wilson, 1991) but little information on the effect of egg weight upon viability or mortality during rearing. Bray (1983a,b), Hearn (1986) and Wilson (1991) have suggested that an integrated broiler operation might find it advantageous to sort eggs by size (with consideration for breeder age), and hatch by egg weight, to give uniform size and prevent dehydration of early hatching chicks. However, experience indicates that Pekin ducklings hatched from eggs <70 g are too small at day-old to be viable in a commercial operation and early hatching day-olds from small eggs often become dehydrated, fail to thrive and die at about 3 days of age.

Factors Affecting Optimum Age at Sexual Maturity

Figure 6.48 shows the relationship between age, the percentage of eggs produced suitable for incubation and hatchability, and the combined effect of these two factors on the number of viable ducklings produced per 100 eggs laid. The graph illustrates the poor productivity experienced prior to 25 weeks suggesting there is no economic benefit in hatching egg production before 24 weeks of age, which means there is no point in aiming for 50% lay before about 25–26 weeks of age.

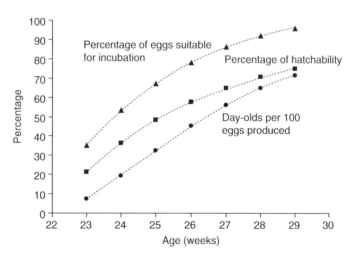

Fig. 6.48. The relationship between Pekin parent flock age and productivity, measured in terms of number of day-olds per 100 eggs produced. Laying performance, egg weight and grading and hatchability based on the results of trials and commercial performance from Pekin parent flocks reared to achieve 80% of *ad libitum* mature live weight at 18 weeks and given a constant daylength of 17 h. Birds were given 2 h feeding time at 20 weeks, increasing by 1 h a week to 6 h at 26 weeks. (Data from Cherry, 1993.)

The optimum economic age at sexual maturity for a particular parent stock genotype is a function of the effect of age at sexual maturity on breeding performance to a fixed age and the growing performance of progeny. Delaying sexual maturity by restricting growth and controlling daylength during rearing increases subsequent rate of lay and egg weight and improves fertility and hatchability of fertile eggs (see Chapter 8, this volume), but there is no economic advantage in delaying sexual maturity for Pekin genotypes with mature weights similar to that of genotypes A and B much beyond about 26 weeks of age.

Sexual Maturity of Egg-laying Breeds

There is little published research into factors affecting age at sexual maturity of ducks used to produce eggs for human consumption, but empirical evidence with Pekin genotypes (live weight about 2.7 kg) selected for egg production both in the EC and in Eastern Europe indicates that egg-laying breeds are affected by the same factors as Pekin selected for growth, body composition and efficiency of feed conversion.

Where duck eggs are sold for human consumption, optimum economic age at sexual maturity will depend on, among other factors, the way eggs are sold. In countries where duck eggs are sold by number or weight, achieving age at sexual maturity early (<22 weeks) is likely to increase economic performance to a fixed age. However, where eggs are sold by grade, the effects of age on egg weight and grade when birds are coming into lay also need to be considered.

Live weight of egg-laying breeds varies between about 1.3 and 2.5 kg, with the heavier breeds used to supply meat as well as eggs. The vast majority of ducks used for the production of eating eggs in the Far East are raised using traditional methods; birds are expected to forage for much of their own feed and this restricts feed intake from an early age. However, attempts to control the spread of avian influenza because of its relevance to human health mean that increasing numbers of ducks are reared semi-intensively, and this is a trend that is likely to continue. Where birds are reared either intensively or semi-intensively, it is necessary to control growth during the rearing of layers in much the same manner as Pekin selected for growth, to prevent premature age at sexual maturity and to achieve optimum laying performance. However, because layers and layer parent stock are much smaller than Pekin meat-type parent stock, they are not able to tolerate the same levels of feed restriction. Empirical experience with Pekin genotypes selected for egg production shows that growth should be controlled to about 70% of *ad libitum* weight by about 35 days and allowed to increase linearly to achieve about 85% of *ad libitum* mature live weight by about 16 weeks of age. Feed allowance should be increased so that by about 18 weeks birds are allowed *ad libitum* feed.

When eggs are required for hatching, the effects of age on egg weight make it preferable to delay age at sexual maturity to about 25 weeks by providing a constant 17 h daylength during rearing, and controlling growth by feed restriction to achieve about 85% of *ad libitum* mature live weight by about 20 weeks of age.

Dietary Regime for Birds on Restricted Feed

Controlling growth can be achieved by restricting feed and thus limiting energy intake (see Figs 6.4 and 6.5) using diets formulated to provide sufficient intake of other nutrients to support limited but healthy growth to maturity. Table 6.9 describes the nutrient composition of starter and grower diets suitable for rearing parent stock and layer ducks on feed restriction from day-old to about 20 weeks of age.

Restricting feed intake may make it necessary, depending upon ingredients, climate, and level of feed restriction, to increase the inclusion of vitamins and trace minerals in the diet. In hot and humid climates, or where feed is likely to be stored for more than about 3 weeks, it may also be necessary to increase the level of supplementation by more than 50% over levels recommended for growing table ducks (see Chapter 5, this volume).

Experience of rearing Pekin ducks with *ad libitum* mature live weight of more than about 4 kg indicates that restricting feed intake below 90 g per bird per day leads to inequitable feed intake, thus increasing variation in live weight and development. Reducing the dietary energy to 11 MJ/kg, compared with the 12–13 MJ/kg, normally provided in least-cost fattening diets in Europe and the USA, provides a practical way of allowing increased feed intake for restricted birds.

It is possible to control growth using either mash or pellet feed and whole grain can be provided up to about 25% of the total feed allowance. Pellet feed is widely used where birds are reared on deep litter or allowed access to free range because pellets can be broadcast over the total floor area, allowing all birds an equal opportunity to obtain a fair share of the available feed. However, to be effective it is essential to use good-quality pellets because although ducks are remarkably adept at finding pellets and whole grain in deep litter, they are unable to find mash fines. Diets for table ducklings are formulated by least-cost methods, which may not necessarily take pellet quality into account. Dust or fines in poor-quality pellets can exceed 10%, affecting both feed intake and growth and substantially increasing feed cost. It is essential to regularly measure

Table 6.9. Nutrient composition of diets suitable for rearing parent stock from day-old to 20 weeks of age.

Nutrient	Starter to 5 weeks	Grower 5–20 weeks
Pellet diameter (mm)	3	4
Energy (MJ/kg)	12	11
Protein (%)	20–22	16–18
Available lysine (%)	1.2	0.80
Available methionine (%)	0.4	0.3
Available methionine and cystine (%)	0.7	0.6
Calcium (%)	0.8–1.0	0.9–1.1
Total phosphorus (%)	0.7–0.8	0.7–0.8
Available phosphorus (%)	0.45–0.55	0.40–0.45

the physical durability of pellet feed (see Chapter 5, this volume) and maintain a pellet durability index >96%.

Mash can be used for reasons of economy or where pellet feed is not available; it is advisable to mix dry mash with water to make a moist but friable mixture. V-shaped feeding troughs are suitable for feeding wet mash, allowing about 12 cm feeding space per bird to permit all birds to feed together.

Husbandry During Controlled Rearing

The systems of housing and environmental control used for rearing parent stock are similar to those used for rearing table duckling (see Chapters 2 and 3, this volume), but controlling growth of parent stock and layers during rearing requires specialized husbandry and management.

Experience shows that rearing parent stock in proximity to other poultry can cause salpingitis, a disease condition where the lumen of the reproductive tract becomes infected and subsequently blocked with purulent material. Salpingitis can cause significant mortality in laying ducks and will substantially reduce laying performance, because chronically affected birds often appear healthy but ovulate into the body cavity. This sometimes causes peritonitis but more frequently yolk material is resorbed and the cycle of ovulation and resorption continues to the completion of lay. To prevent salpingitis and other diseases, it is essential to rear parent stock from day-old in single-age accommodation, which must be clean, properly disinfected (see Chapter 3, this volume) and located well away from other poultry.

Site hygiene must be of a high standard and visitors should be reduced to a minimum. In particular, it is important to prevent anyone who has had recent contact with other poultry from entering the rearing accommodation. All staff and visitors should change into protective clothing and footwear, use footbaths and wash hands with a suitable disinfectant prior to entering the rearing accommodation. Where ducks are allowed access to bathing or swimming water, the water must be provided from a clean source. It is important to obtain competent veterinary advice regarding vaccination and to make arrangements to obtain supplies of vaccines from a competent, government-approved source and to vaccinate as required by local legislation.

Controlling growth delays normal physiological development. When rearing parent stock or layers on feed restriction from day-old, it is important to relate husbandry and environment to the needs of the bird (see Chapter 3, this volume) and not to chronological age. Behaviour provides a good indication of the birds' comfort. Feed restriction affects growth and development of down (see Fig. 6.3), delaying the age at which birds might be expected to control their body temperature. Breeder chicks on restricted feed require brooding with supplementary heat to a later age than table ducklings.

Where parent stock and layers are reared intensively on litter to about 20 weeks it is necessary to provide about 0.45 m² space per bird. This provides sufficient space to maintain litter quality and, where pellet feed or whole grain is broadcast over the litter area, allows all birds to feed together and obtain a fair share of the available feed. Initial floor space can be reduced in cold climates to reduce heating costs by using plastic curtains or an insulated partition to limit the brooding area to about $0.25\,m^2$ per bird. When heat is no longer required, birds can be allowed access to the total floor area. Where parent stock are reared semi-intensively, they require about $0.25\,m^2$ per bird of covered accommodation with access to about $0.5\,m^2$ of outside pen space, roofed to provide protection from rain, snow and sun. It is important to provide bird-proof fencing around the outside of pens and swimming channels to prevent predation and contact with wildfowl, which frequently carry disease that can affect domestic duck.

To maintain birds in good health and feather condition the entire floor area should be covered daily with fresh clean litter. Materials such as straw, shavings and rice hulls are all suitable bedding materials but ducks are susceptible to fungal infections so it is important to use only clean and dry litter, which should be stored under cover.

Drinkers must be located over a suitably drained area and a slatted area should be provided adjoining any drinking or bathing water (see Chapter 3, this volume) to prevent birds splashing water on to littered areas. Many countries have enacted vigorous environmental protection legislation to prevent pollution of water supplies. Within the EC, the USA and Canada effluent from intensive and semi-intensive rearing facilities must be either treated and discharged into approved drainage facilities, or irrigated over land in such a manner as to prevent any runoff into streams or watercourses.

When controlling growth of meat-type parent stock, it is sometimes necessary to rear drakes from genotypes selected for rapid growth separately from ducks of slightly slower-growing genotypes. However, rearing drakes in single-sex pens can affect subsequent mating behaviour at maturity, with substantial numbers of males preferring to mate with other males, thus reducing fertility and overall breeding performance. If the sexes are reared separately, it is important to 'sexually imprint' drakes, by rearing them with a small number of ducks from day-old; a ratio of about one duck to every eight drakes is sufficient to ensure normal mating behaviour.

To achieve optimum rate of lay it is essential to begin feed restriction from day-old. Figure 6.49 shows a pattern of controlled growth as it should be for rearing meat-type Pekin parent stock to maturity. The level of feed restriction required to achieve this controlled growth depends upon genotype, climate and dietary energy (see Figs 6.7 and 6.8). To achieve this pattern of growth in temperate climates, using suitable diets (see Table 6.9), it is necessary to restrict feed consumption to about 50–60% of *ad libitum* feed.

To achieve full control of growth, live weight must be monitored at 2-week intervals from about 3 weeks of age, adjusting feed allowance as necessary to maintain the required rate of gain. A 15% sample is sufficient to provide a reasonable estimate of flock live weight. Growth charts along with statistical process control (see Chapter 4, this volume) provide a good method of monitoring live weight over time.

At day-old, the feed allowance should be spread on suitably sized trays located close to both heat and water. As birds get older, increasing amounts of feed

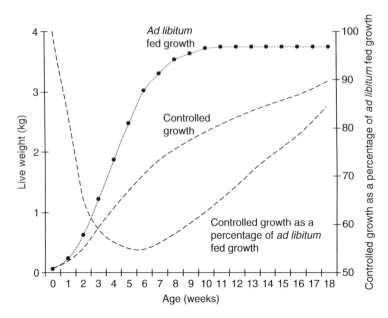

Fig. 6.49. Controlled growth of Pekin duck to maturity expressed as a percentage of *ad libitum* growth. (Data from Cherry, 1993.)

should be broadcast on the littered area. Birds consume their feed allowance in about 6–8 h at day-old, declining to 1 h by 21 days and 10–15 min by 5 weeks.

To allow all birds an equal opportunity to get a fair share of the feed it is important to spread the pellets over a wide area of the floor as quickly as possible. To limit competition and allow for rapid distribution of feed it is sensible to limit pen size to about 300 birds. Wet mash can be fed in V-shaped troughs or spread on the ground where birds have access to outside pens. There should be a constant supply of drinking water, using either pig troughs, bell-shaped turkey drinkers or nipple drinkers (see Chapter 3, this volume). It is essential to supply sufficient drinkers to enable all birds to drink at the same time after feeding.

It is important to monitor growth, appearance and health by inspecting birds frequently from day-old to 5 weeks and then at 2-week intervals to maturity. To be effective it is essential to carry out these supervisory inspections using some form of checklist to examine facilities, environment, physical feed quality and feeding behaviour, bird feathering and health in an objective manner.

References

Assenmacher, I. (1974) External and internal components of the mechanism controlling reproductive cycles in drakes. In: Pengelly, E.T. (ed.) *Circannual Clocks*. Academic Press, London, pp. 197–248.

Bray, T. (1983a) Broiler chick quality problems. *Gleadthorpe Experimental Husbandry Farm Booklet* 10, 4–9.

Bray, T. (1983b) Broiler chick weight, does it matter? *Gleadthorpe Experimental Husbandry Farm Booklet* 11, 17–20.

Cardinali, D.P., Cuello, A.E., Tamezzani, J.H. and Rosner, J.M. (1971) Effects of pinealectomy on testicular function of the adult male duck. *Endocrinology* 89, 1082–1093.

Cherry, P. (1993) Sexual maturity in the domestic duck. PhD thesis, University of Reading, Reading, UK.

Chueng-Shyang Ma, R. (1968) The time of release of the luteinizing hormone from the adenohypophysis of laying domestic duck. *Poultry Science* 47, 404–409.

Gilbert, A.B. (1979) Female genital organs. In: King, A.S. and McLelland, J. (eds) *Form and Function in Birds*. Academic Press, London, pp. 295–301.

Gille, U. and Salomon, F.V. (1998) Muscle growth in wild and domestic ducks. *British Poultry Science* 39, 500–505.

Hearn, P.J. (1986) Making use of small hatching eggs in an integrated broiler company. *British Poultry Science* 27, 498.

Hocking, P.M. (1990) Comparison of the effects of the degree of food restriction on ovarian function at the onset of lay in unselected ducks (Anas platyrhynchos), and in a line selected for improved food efficiency. *British Poultry Science* 31, 351–359.

Hollands, K.G. and Gowe, R.S. (1961) The effects of restricted and full feeding during confinement rearing on first and second year laying performance. *Poultry Science* 40, 574.

Krapu, G.L. (1981) The role of nutrient reserves in Mallard reproduction. *Auk* 98, 29–38.

Olver, M.D. (1984a) Qualitative versus quantitative feed restriction in Pekin breeder ducks during the rearing period. *South African Journal of Animal Science* 14, 75–78.

Olver, M.D. (1984b) Quantitative feed restriction in Pekin breeder ducks during the rearing period and its effect on subsequent productivity. *South African Journal of Animal Science* 14, 136–141.

Olver, M.D. (1986) Performance of Pekin ducks subjected to qualitative feed restriction at various ages during the rearing period. *South African Journal of Animal Science* 16, 43–46.

Olver, M.D. (1988) Quantitative feed restriction of Pekin ducks from 3 weeks of age and its effect on subsequent productivity. *South African Journal of Animal Science* 18, 93–96.

Olver, M.D., Kuyper, M.A. and Mould, D.J. (1978) Restricted feeding of Pekin breeder ducks during the rearing period and its effects on subsequent productivity. *Agroanomalia* 10, 7–12.

Pattenden, R.K. and Boag, D.A. (1989) The effects of body mass on courtship, pairing and reproduction in captive Mallards. *Canadian Journal of Zoology* 67, 495–501.

Poole, E.L. (1938) Weights and wing areas of North American birds. *Auk* 55, 511–517.

Prince, H.H. (1979) Bioenergetics of post breeding dabbling ducks. In: Bookhaut, T.D. (ed.) *Waterfowl and Wetlands – An Integrated Review, Proceedings of Symposium 39th Midwest Fish & Wildlife Conference*. The Wildlife Society, Madison, Wisconsin, pp. 103–117.

Shanawany, M.M. (1987) Hatching weight in relation to egg weight in domestic birds. *World's Poultry Science Journal* 43, 107–115.

Sharpe, P.J. (1984a) Seasonal breeding and sexual maturation. In: Cunningham, F.J., Lake, P.E. and Hewitt. D. (eds) *Reproductive Biology of Poultry*. British Poultry Science, Longmans, Edinburgh, UK, pp. 203–218.

Sharpe, P.J. (1984b) Seasonality and autonomous reproductive activity in birds. *Bolleettino di Zoologia* 51, 345–403.

Wilson, H.R. (1991) Interrelationships of egg size, chick size, post-hatching growth and hatchability. *World's Poultry Science Journal* 47, 5–20.

7 Management of Breeding Ducks

As noted in Chapter 2 (this volume), there is a great variety of systems in use around the world for keeping breeding ducks. This means that descriptions of daily routine are not appropriate, but there are some points about managing breeding ducks that are general and will apply in most, if not all, situations.

Table 7.1 lists laying performances recorded for a variety of locations and genotypes, and shows the remarkable productivity achieved by the domestic duck maintained under a wide range of husbandry systems.

Environment

High temperature reduces feed intake and rate of lay of both parent stock and egg layers. Figure 7.1 describes the effect of temperature on feed intake of several commercial Pekin breeding flocks with a mature live weight of about 3.6 kg, laying about 5 eggs per female per week with a mating ratio of about 6 ducks per drake and provided with a 10.8 MJ/kg diet. Analysis of these data provides the following estimate of the effect of ambient temperature (x, °C) upon feed intake:

Feed intake (g/day) = $237.47 - 0.36x - 0.0589x^2$

The relationship between environmental temperature and feed intake is curvilinear but declines by about 2.2 g (24 kJ) per degree increase in temperature in the range from 5°C to 25°C. However, genotype, live weight (see Fig. 5.11), dietary energy, rate of lay, provision of swimming water and acclimatization will all affect energy intake.

There are no reports of controlled experiments on the effect of temperature on nutrient intake and consequent laying performance. However, records of commercial performance confirm that rate of lay is not affected by average ambient temperatures as high as about 24°C provided the intake of nutrients is

Table 7.1. Effect of location, genotype and production system on the laying performance of domestic duck.

Location	System	Breed	Mature weight (kg)	Laying period (weeks)	Eggs per female	Egg weight (g)
Far East[a]	Herded	Khaki Campbell	1.8	52	271	71
	Confined	Khaki Campbell	1.8	52	242	70
Far East[b]	Duck-cum-fish	Khaki Campbell	1.5	52	255	61
		Indian Runner	1.6	52	248	61
		Zending	1.5	52	223	61
Far East[c]	Herded	Khaki Campbell	1.6	52	267	67
		Khaki Campbell	1.7	52	279	68
Far East[d]	Intensive	Pekin	3.2	37	185	86
Far East[c]	Confined	Triet Gang	1.5	52	255	58
		CV 2000	1.8	52	275	67
		Co	1.8	52	225	62
		Pekin	2.7	40	195	78
		Bau	1.8	40	120	78
		Ky Lua	1.8	40	135	74
E. Europe[d]	Intensive	Pekin	3.0	40	226	88
E. Europe[d]	Semi-intensive	Pekin	3.8	30	168	85
UK[d]	Intensive	Pekin	3.5	40	218	88
		Pekin	4.0	40	175	90
USA[d]	Intensive	Pekin	4.0	40	215	87
France[e]	Intensive	Pekin	na	50	150	na
		Tsaiya	na	50	214	na
Holland[f]	Semi-intensive	Khaki Campbell	1.6	52	335	73.4

[a]Far East, Nho and Tieu (1996); [b]Far East, Das *et al.* (2003); [c]Far East, Tran Thanh Van, Thai Nguyen University, Vietnam, 2005, personal communication; [d]Far East, E. Europe, UK and USA, Cherry (1993); [e]France, Valez *et al.* (1996); [f]Holland, Hutt (1952).

maintained. Like laying fowl, the breeding duck's requirement for nutrients other than energy is independent of temperature. Trials show that there is a curvilinear relationship between protein intake and rate of lay, which suggests that Pekin parent females require about 42 g of protein, providing not less than 2 g lysine and 1.4 g methionine plus cystine daily, to maintain maximum rate of lay.

Increasing ambient temperature reduces feed and nutrient intake and this begins to affect rate of lay when average ambient temperature exceeds about 24–25°C. This is the point at which laying ducks (depending on bodyweight, feathering, composted litter temperature, access to swimming water and relative

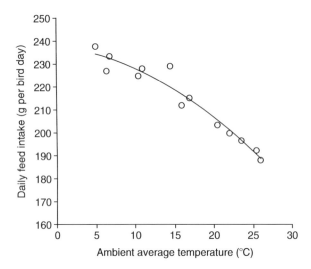

Fig. 7.1. The relationship between average ambient temperature and feed intake per bird recorded for each of 13 lunar months for Pekin parent stock with a mature live weight of 3.6 kg, given about 9 h access per day to a 10.8 MJ/kg diet.

humidity) start panting, using latent heat exchange via respiration to achieve sufficient heat loss to maintain homeostasis. Trials with Pekin parent stock fed *ad libitum* and maintained at average ambient temperatures above 25°C showed that increasing the intake of essential amino acids by raising dietary protein above the levels described in Table 7.2 did not improve rate of lay. This is because energy intake, not amino acid supply, limits performance at high temperatures. However, attempts to increase energy intake by increasing dietary energy merely depressed feed intake and did not enable birds to increase their intake of energy. Attempts to encourage feed intake by providing wet mash or using whole grain as a scratch feed or by allowing birds to feed at night did not increase feed intake for more than a few days, indicating that birds control their feed intake based on their requirement for energy, but only as far as this is consistent with maintaining normal body temperature.

Giving birds an environment where they can be comfortable is essential to achieving optimum breeding performance. Ducks can tolerate a wide range of temperature but mature birds appear comfortable between about 6°C and 23°C. As temperature falls towards freezing, birds reduce sensible heat loss by drawing their legs and feet under their body and placing their bill under their wing. They also congregate in 'rafts' resting on litter kept warm by fermentation.

Heat from deep litter fermentation is widely exploited in cold climates and during winter in temperate climates to maintain shed temperature above 6°C. Duck effluent, water splashed about when ducks drink and enthusiastically preen themselves, along with fresh straw, shavings or rice hulls provided daily to keep the litter in good condition, encourage fermentation so that litter maintains a steady temperature of about 30°C. In Eastern Europe and Russia, litter fermentation is used to provide enough heat to maintain temperatures above

Table 7.2. Nutrient composition of a diet suitable for breeding and laying duck.

Nutrient	Breeder fed from 20 weeks
Pellet diameter (mm)	5
Energy (MJ/kg)	10.8–11.2
Protein (%)	20.5
Available lysine (%)	0.97
Available methionine	0.31
Available methionine and cystine (%)	0.73
Calcium (%)	3.21
Total phosphorus (%)	0.6
Available phosphorus	0.31

freezing when outside temperatures are below −15°C. Birds remain comfortable, exhibiting normal behaviour even when inside air temperature falls below freezing because they are resting on a warm littered floor. However, to sustain fermentation and good litter condition, which is essential for feathering, bird health and breeding performance, it is necessary to control ventilation to maintain air quality but reduce heat loss to a minimum.

When temperature rises beyond about 23°C, birds expose and extend their legs and wings and open their feathers to encourage sensible heat loss; they become increasingly lethargic and, depending on relative humidity, start panting to control body temperature. Birds avoid resting on warm litter, preferring slatted flooring or shaded areas in outside pens which allow them to attain maximum sensible heat loss. Slats can be used for more than 35% of the floor area for intensively housed birds in hot climates. Removing litter regularly prevents litter fermentation and reduces litter temperature, but replacing litter must be done regularly from the time birds are housed to allow birds to become accustomed to this activity.

Ducks spend a lot of time drinking, preening and socializing, and as environmental temperature rises above the panting threshold of 24–25°C, birds spend increasing amounts of time resting and loafing around, usually close to water. Birds avoid resting in direct sunlight where radiant temperature can exceed 50°C. Where drinking and bathing water is located down one side of the accommodation it is sensible to extend roof eaves or to provide blinds to enable birds to drink in the shade. To encourage birds to use outside pens and water provided for bathing and swimming, it is essential to provide shade to allow birds to avoid high radiant temperature.

Commercial trials investigating the effect of evaporative cooling on performance of Pekin parent stock maintained at high ambient temperature showed that it is possible to reduce average day- and night-time dry bulb temperatures by about 5°C and 1°C, respectively (see Chapter 4, this volume). Fogging the atmosphere was also effective in reducing dry bulb temperature by as much as 8°C. However, neither method of cooling had any measurable effect on feed intake because, with both systems, the reduction in dry bulb temperature could only be achieved at the expense of an increase in relative humidity. This reduces the birds' ability to achieve heat loss by vaporization of moisture from the lungs and trachea when panting. For this reason, reducing dry bulb temperature by fogging did not reduce 'effective temperature'. Also, litter condition deteriorated, which affected feathering, bird appearance and egg cleanliness.

Commercial-scale trials using tunnel ventilation, overhead fans or pressure jets to modify air speed for parent stock experiencing high temperatures showed that, as house temperature increased towards midday, birds preferred resting either close to air inlets or directly underneath overhead fans and pressure jets. However, although birds clearly preferred resting in a stream of fast-moving air, the treatments had no measurable effect on feed intake or laying performance.

Experience in Eastern Europe, which enjoys high summer temperatures and where parent stock are allowed access to fish ponds, and from the Far East shows that providing bathing water can help maintain laying performance at high ambient temperature. However, legislation, and increasing public health concerns regarding the discharge of water effluent limit the opportunities to provide either bathing or swimming water in the European Community (EC) and North America.

Nutrition

There is considerable variation in the estimates of the nutrient requirements of breeding and laying ducks provided by Shen (1985), Dean (1985), Elkins (1987) and Scott and Dean (1991), and the recommendations provided by the Agricultural Research Council in the UK, the National Research Council in the USA and commercial breeding companies. Estimates for the requirement for protein and first limiting amino acids, lysine and methionine, plus cystine vary from 15% to 19.5%, 0.6% to 1.1% and 0.5% to 0.68%, respectively. This variation arises partly from differences in genotype and climate, but mainly from differing methods used to determine 'requirement'.

Data from a trial by Pan *et al.* (1981), as reported by Shen (1985), show a curvilinear relationship between increasing protein in the range of 15–21%, at two dietary energy levels, on rate of lay and egg weight, with rate of lay increasing up to 19% protein at both energy levels (see Fig. 7.2). Birds given the 11.08 MJ/kg diet at the lower concentrations of protein achieved a greater egg weight probably because, although feed intake was not reported, the birds increased their feed intake to satisfy their requirement for energy, thus increasing their intake of protein. Analysis of data from unreported trials (see Figs 7.3 and 7.4) shows a curvilinear relationship between increasing daily protein intake in the range of 30–50 g on both egg mass and egg weight.

Table 7.2 gives the nutrient composition of a diet which will allow commercial Pekin parent stock to achieve about 225 eggs in 40 weeks of lay, although these values should not be regarded as 'requirements' because both rate of lay and egg weight respond in a curvilinear manner to increasing nutrient concentration. This means that it is necessary to relate the cost of nutrient specification to the

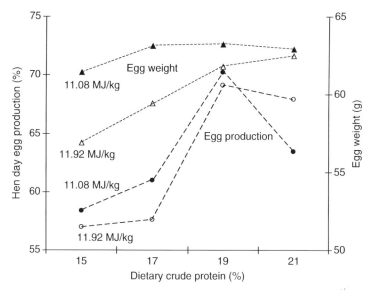

Fig. 7.2. Effect of dietary protein and metabolizable energy on laying performance and egg weight over a 44-week laying period of *ad libitum* fed white Tsaiya given maize–soybean diets containing graded levels of protein at two levels of ME. (Data from Shen, 1985.)

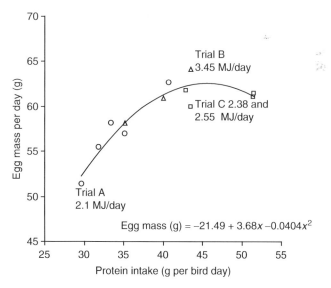

Fig. 7.3. Relationship between protein intake and egg mass per day recorded between 34 and 42 weeks of age. In three separate trials genotype B ducks were given fixed quantities of feed providing increasing increments of protein, and all treatments within each trial received the same daily intake of energy.

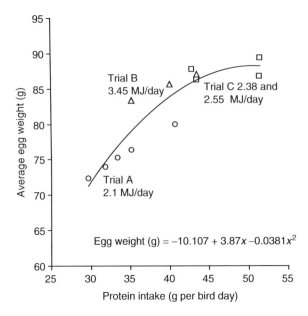

Fig. 7.4. Relationship between protein intake and average egg weight recorded between 34 and 42 weeks of age. Genotype and treatments as in Fig. 7.3.

expected benefit, which is affected by genotype, feed intake and the value of eggs for hatching or eating. For example, Pekin parent stock and Khaki Campbell layers produce a similar daily egg mass of between 60 and 65 g per bird and, despite differences in bodyweight, it is probable that their response to increasing increments of essential amino acids is similar. However, the value of their eggs affects the cost/benefit ratio of increasing nutrient density, explaining why Pekin parent stock are usually provided with a substantially higher protein diet than Khaki Campbell layers.

Controlling Egg Weight

Controlled feed intake, using either quantitative restriction or controlled feeding time, can be employed in Pekin genotypes selected for rapid growth to control egg weight and improve hatchability (see Chapter 8, this volume). Analysis of trials using Pekin ducks with *ad libitum* fed live weight between 4 and 5 kg and diets containing about 10.8 MJ/kg diet shows a curvilinear relationship between feed intake and egg weight (see Fig. 7.5). The data provide the following estimate of the effect of feed intake (x, g per day) on egg weight:

$$\text{Egg weight (g)} = -21.169 + 0.841x - 0.0016x^2$$

Reducing feed intake by 12 g or 130 kJ per day reduced egg weight by about 2 g on average over the range, but genotype, mature weight and dietary energy can alter the pattern of the response.

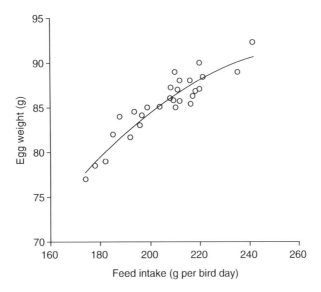

Fig. 7.5. Relationship between controlled feed intake and average egg weight between 34 and 40 weeks of age for Pekin parent stock with a mature weight of about 3.8 kg given a diet containing 10.8 MJ/kg and 20% protein.

The feed programme provided when birds are coming into lay can affect age at sexual maturity, rate of lay and egg weight. Many producers rigorously control feed intake when birds are coming into lay in the mistaken belief that if birds are given increased feed the gain in weight will adversely affect subsequent rate of lay. However, Fig. 7.6 shows that birds given increasing feed from 20 weeks and provided with timed feeding from 22 weeks did not gain much weight, but gave increases in egg size not far below those of the 95% live weight treatment. On the other hand, Fig. 7.7 shows that rigorously limiting feed intake when birds are coming into lay substantially reduces both rate of lay and egg weight. To achieve optimum laying performance to a fixed age for Pekin parent stock housed in temperate climates, it is necessary to increase the daily feed allowance from about 18 weeks of age by 15 g per bird each week, or to provide 2 h feeding time per day at 18 weeks of age and increase this weekly to between 6 and 8 h access to feed by 21–23 weeks of age.

To achieve good hatchability in temperate climates from genotypes selected for rapid growth, egg weight must be limited to about 93 g by restricting energy intake through either quantitative or timed feeding. Age affects egg weight (see Fig. 7.6) and it is important to encourage feed intake when birds are coming into lay to achieve maximum egg weight in the early stages. Experience shows that applying feed restriction to genotypes selected for rapid growth to control egg weight after about 30 weeks of age can sometimes affect post-peak laying performance. A sensible approach is to provide very moderate feed restriction by allowing birds between 6 and 8 h feeding time from about 23 weeks and

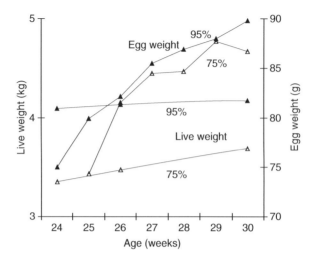

Fig. 7.6. Relationship between age, live weight and egg weight for genotype B reared to achieve 75% or 95% of *ad libitum* fed live weight at 18 weeks. Birds fed to achieve 75% of mature weight were provided with 7 h timed feed restriction from 22 weeks; the other rearing treatment continued on *ad libitum* feed. Growth and light programme as in Fig. 6.19 and laying performance as in Fig. 6.20. (Data from Cherry, 1993.)

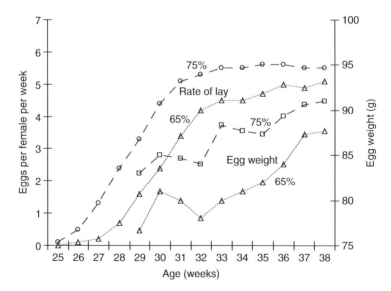

Fig. 7.7. Effect of controlled feed intake on subsequent rate of lay and egg weight of genotype B Pekin females reared to achieve 65% or 75% of *ad libitum* fed mature live weight at 18 weeks and then maintained at the same body weight in lay by quantitative restriction. Birds were given 145 or 190 g from 25 to 32 weeks and 190 or 220 g from 33 to 38 weeks of a diet containing 10.8 MJ/kg and 21% protein. (Data from Cherry, 1993.)

then to gradually reduce feeding time or feed quantity as egg weight approaches 90 g. However, it is also necessary to take environmental temperature into account because increasing temperature also reduces feed intake (see Fig. 7.1) and subsequent egg weight (see Fig. 7.5), which means that during the summer birds must be allowed more feeding time when mean of daily maximum and minimum ambient temperature increases above 18°C. Above about 24°C feed should be provided *ad libitum*.

Unpublished trial results show that reducing feeding time for Pekin parent stock reduces feed intake in a curvilinear manner (Fig. 7.8), with an 8% reduction for halving feeding time and a further 8% for halving it again. To control egg weight by restricting energy intake but to maintain laying performance, it is essential to provide a diet formulated to allow birds to achieve their requirement for other nutrients. When formulating duck breeder diets it is desirable to know both feed intake and environmental temperature. The nutrient composition described in Table 7.2 is sufficient to allow Pekin females with a mature weight of about 3.6 kg, given 200–210 g feed per bird per day, to maintain >6.2 eggs per female per week with an average egg weight of between 88 and 92 g.

It is important to sample and record egg weight every 2 weeks and to adjust feeding time or quantity so as to maintain control of egg weight. It is sensible to use process control charts (see Fig. 4.44) that show the relative rate of change in egg weight, and to alter daily feed allowance or feeding time to prevent egg weight increasing beyond about 92 g. However, controlling egg weight is as much a skill as a science, requiring experience in the effects of genotype, age, environmental temperature and nutrient composition upon food intake, rate of lay and egg weight.

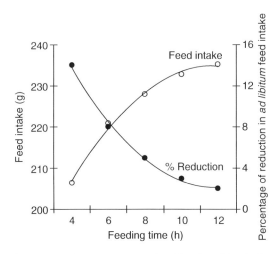

Fig. 7.8. Effect of controlled feeding time on feed intake of Pekin females with a mature weight of 3.6 kg laying about 6 eggs per female per week and given a diet containing 10.8 MJ/kg.

Housing

In tropical regions of the Far East and China, parent stock are provided with open-sided housing and frequently given access to bathing water in outside pens. Houses are usually about 12 m wide and 3 m to the eaves, with roofing insulated with natural materials and pitched at 30–40° to encourage natural ventilation by 'stack effect'. In tropical and subtropical latitudes, orienting the long axis of the house in an east–west direction and allowing the roof to over-hang sidewalls by about 1.75 m can prevent direct sunlight from entering the open sidewalls. Ducks choose to avoid direct sunlight in hot weather and providing shade allows them to rest and bathe outside during the heat of the day.

Slats are widely used because they provide a ventilated and relatively cool surface for birds to rest on, but to avoid foot problems with Pekin parent stock it is necessary to limit the slatted area to about 35% of the floor area. Bamboo is widely used as slatted flooring, but should not be fixed to joists with nails because bamboo splinters easily, causing foot infections, lameness and substantial adult mortality. Where birds are confined within covered accommodation at night it is necessary to provide drinking water using nipple drinkers, plastic turkey drinkers, pig troughs (see Chapter 4, this volume) or open troughs. In wet tropical, Mediterranean and warm temperate climates birds are given similar accommodation, but solid end walls, dwarf sidewalls and sidewall curtains are used to provide protection against occasional strong winds and driving rain.

In Eastern Europe, which enjoys hot summers and cold winters, parent stock are provided with well-insulated accommodation and are allowed access to outside pens during daylight hours in summer, and when conditions permit in winter. As noted above, litter fermentation is encouraged by reducing space allowance per bird (0.25 m² per bird) and providing ample fresh litter daily. Ventilation is controlled and the heat from litter fermentation is sufficient to maintain a comfortable temperature even during sustained periods of very cold weather.

In cool temperate and cold climates, increasing labour costs and concerns about risks from transmission of disease from wild birds, particularly migrating wildfowl, means that most parent stock in the USA, EC and imported geno-types in the Far East and China are housed intensively, using similar housing to that provided for other poultry (see Chapter 3, this volume).

Birds housed intensively without access to outside pens require about 0.45 m² space per bird, with litter providing about 80% of the floor area to supply sufficient heat to maintain temperature above 8°C, the minimum required in winter to maintain litter in a relatively dry and clean condition. Drinkers should be placed over a drainage pit located at least 0.75 m above the floor and covered with either pig-type plastic slats or plastic-coated weld-mesh. Where legislation prevents the discharge of duck effluent it is neces-sary to eliminate slatted flooring and provide water through nipple drinkers located over litter.

Furniture and Equipment

To prevent the transmission of disease, both staff and visitors must wash their hands and change into suitable protective clothing and footwear before entering breeder and layer accommodation. The evidence from trials shows that following these simple procedures before entering each separate breeder or layer unit can substantially reduce the incidence of 'egg drop syndrome' and similar diseases. There should be a small vestibule with facilities for hand washing, and changing footwear and outer clothing at the entrance of each breeder shed.

Figure 7.9 shows a nest design with individual nest dimensions of 400 × 400 mm, suitable for Pekin parent stock. Trials investigating the effect of nest design on laying behaviour found that birds are unwilling to use tiered nests, but show no preference for any particular shape. Nest design had no substantial effect on incidence of cracked or floor eggs. It is convenient to store litter used both for bedding and as nesting material inside the breeding accommodation on tables located close to sidewall doors to assist delivery and reduce handling.

Drinkers should be placed over a slatted drainage channel with effluent collected for subsequent irrigation over arable land. Where regulations or circumstances prevent irrigation, water can be provided by nipple drinkers located over litter to absorb waste water. Nipple systems should be installed so that they can be moved at regular intervals to maintain litter in a clean and dry condition. It is important to control water pressure to that recommended by the manufacturer.

Fig. 7.9. Nest design suitable for parent stock with a mature live weight of about 3.6 kg.

The feeder design described in Fig. 3.20 is suitable for adult birds fed *ad libitum* in tropical and Mediterranean climates, but in cold and cool temperate climates it is sometimes necessary to place flaps over feed troughs by hand or by using a time switch to control access to feed.

It is recommended that both breeding and laying ducks are given a 17 h daylength using either fluorescent or incandescent electric light or paraffin lamps to provide a minimum light intensity of 5 lx, sufficient to read the text of a newspaper at normal reading distance. Where birds are housed in open-sided housing it can sometimes be necessary to provide screening to prevent car headlights and shadows created by moonlight from alarming birds.

Preparing Accommodation for Parent Stock or Laying Ducks

It is important to clean thoroughly and disinfect parent and laying stock accommodation, feed and water supply systems, and to clean, service and check all equipment, electrical and mechanical systems and components for safety (see Chapter 3, this volume). Before the arrival of birds it is important to check and adjust as necessary both water and gas pressure and distribute about 100 mm of clean dry litter over the floor and check that all equipment is working and adjusted correctly.

Nests at the ratio of 1 for every 4 or 5 birds should be located along all sides of each pen, or sidewalls and end walls where birds are run as one flock. About 120 mm of clean and dry litter (preferably wood shavings) should be provided in each nest.

Ventilation should be set up and monitored as described in Chapter 3 (this volume). Finally, it is advisable to carry out a careful inspection using a checklist to make sure the accommodation and equipment are prepared and set up for the arrival of the birds.

Transporting and Housing Parent Stock and Laying Ducks

Females and drakes along with 'imprinting' females should be moved to their new accommodation early in the day to enable them to 'find water' before being fed, and to become familiar with their new surroundings before nightfall. Transporting ducks requires care and kindness. By far the simplest method to move ducks is to walk birds up a ramp onto a suitably designed low trailer for transport to their new accommodation. It is possible to transport birds long distances in this manner without causing distress, and they can be walked off the trailer into their new accommodation.

It is important to count the birds accurately and to provide sufficient drakes to achieve a mating ratio of 1 drake to 5 or 6 ducks for Pekin parent stock or 1 drake to 6 or 7 ducks in the case of laying strains.

Ducks frequently have difficulty in 'finding' water when moved into new accommodation. It is worthwhile penning birds in close proximity to drinkers until it becomes obvious that birds are drinking and preening feathers normally; with nipple drinkers it is necessary to arrange for some nipples to drip continuously until birds find the water.

Birds should be given breeder feed (*see* Table 7.2) at about 20 weeks of age, just prior to the onset of lay, and if birds are to be maintained on feed restriction they should be given a liberal supply of feed for 2 or 3 days after moving. In temperate and cool climates both Pekin parent stock and laying duck should be provided with at least 5 h feeding time a day when first housed. In the tropics birds can be fed *ad libitum*. However, to prevent birds choking on dry feed, they should not be given feed on the day they are moved until they have settled into their new accommodation, and then only when they have found the water and are seen to be drinking and preening normally.

Routine Husbandry of Parent Stock and Laying Ducks

Egg formation takes about 24 h from ovulation to oviposition. Ducks lay their eggs early in the morning commencing at daybreak or when lights come on, whichever is the earlier. Birds bury their eggs in the litter, and when collecting eggs it is necessary to sift carefully through the litter in each nest to find all the eggs; usually only one or two collections are required. Eggs should be collected onto stackable plastic turkey egg trays designed to be self-supporting to prevent damage to the eggs. Dirty and cracked eggs should be placed on to a separate tray.

Ducks often lay up to 30–40% of their eggs on the floor in the early stages of lay and even apparently clean floor eggs are usually heavily contaminated, affecting subsequent hatchability. Attracting birds to use nests depends on nest location and on providing relatively soft, dry and friable nest litter such as wood shavings or rice hulls, materials that birds can bury their eggs in. If possible, a nest litter of different appearance from that used for general bedding should be provided. To produce clean eggs it is important to persuade ducks to use nests, to prevent access to nests after the last egg collection until nightfall and to supply fresh clean litter to all nests daily in the late afternoon. Accumulated nest litter should be removed and spread over the floor at weekly intervals, but it is important to carry out this routine in the afternoon. Disturbing birds early in the morning reduces egg production the following day by about 5–8%, because birds are ovulating at this time and any disturbance can cause them to ovulate into the body cavity.

It is essential to carry out a daily inspection of birds and accommodation, looking in particular at bird appearance and behaviour, litter and air quality (*see* Chapter 4, this volume). Litter quality affects feathering, behaviour, bird health and breeding performance. It is important to spread fresh clean litter over the floor area every day, and to maintain shed temperature above 8°C and relative humidity beneath 80% to assist fermentation.

All drinkers and swimming channels should be drained, thoroughly cleaned and refilled daily. The feed supply should be checked and feeders inspected to make sure that both mechanical and control systems are working properly so that birds have an adequate supply of feed.

Recording daily maximum and minimum temperatures, water use and feed intake for birds given either *ad libitum* or timed quantitative feed restriction can sometimes provide early warning of impending health problems. Feed records can also be useful when investigating problems associated with laying performance and hatchability. Egg production, quality and hygiene should be recorded daily. If daily egg production declines by more than about 10% the person responsible for day-to-day husbandry should inform the supervisor.

Weekly Routines

Egg weight should be monitored weekly by weighing a random sample of 100 eggs from each breeding flock. Egg weight along with other performance data including rate of lay, egg hygiene, fertility and hatchability should be compared with planned performance. All breeding and laying flock locations should be inspected weekly by a trained supervisor using a checklist to carefully inspect and record the condition of the accommodation, litter, nest hygiene and air quality. It is important to take time to examine bird appearance, behaviour, mobility and health, and check control gear and operation of ventilation, lighting and mechanical equipment, and both site and housing for signs of vermin. Immediate problems should be discussed and if possible remedied on site with the person responsible for the day-to-day husbandry.

Companies with several breeding or laying flocks should consider monitoring performance using 'statistical process control' (see Chapter 4, this volume) to provide warning when either a common cause or a special cause of variation is affecting performance.

Second Lay

When Pekin parent stock have completed >40 weeks in lay, rate of lay and hatchability decline to a point where it is necessary either to replace the flock or to moult the birds and continue into a second lay.

The moult in domestic duck is influenced by its inheritance from its wild forebear and moulting is stimulated in mature wild Mallard by the declining days following the summer solstice. Birds lose all their primary feathers more or less simultaneously and temporarily lose the ability to fly. The moult is sometimes described as the 'eclipse moult' because drakes lose their distinctive plumage and become indistinguishable from females. During the moult, drakes are completely infertile. The moult is complete within 6–8 weeks, in time for the annual migration in the autumn or fall.

Moulting in domestic duck is no longer controlled by decreasing daylength, but birds still retain the inherited characteristic of moulting all their primary feathers simultaneously. Birds given adequate nutrition can complete their moult and commence lay within about 7–8 weeks. To stimulate moulting it is helpful to reduce daylength to about 8 h and when natural daylength exceeds about 10 h it may be necessary to temporarily move birds housed in open-sided or windowed accommodation to controlled-light accommodation to provide a short daylength of 8 h. In addition to reducing daylength it is necessary to restrict feed intake to maintenance requirement (see Fig. 5.11). After about 10–14 days or when the majority of birds are moulting, feed allowance should be increased to provide sufficient nutrition for replacing primary feathers. Birds should be given at least 2 h feeding time, increasing by an hour a week to about 8 h, or *ad libitum* feed at high ambient temperature. Daylength should be increased by an hour a week to 17 h. It is important to note that moulting and second-lay performance do not require or depend on weight loss; domestic ducks, unlike hens, do not gain much weight in lay, and attempts to reduce bodyweight by maintaining birds on less than maintenance energy requirement delays replacement of primary feathers, and substantially reduces subsequent breeding performance.

Laying performance of second-lay birds follows a similar pattern to first lay, except for egg size, which returns immediately to the pre-moult level. Fertility and hatchability are substantially improved following the moult. Rate of lay is restored, but declines at a faster rate than during first lay, and after about 25–30 weeks it is normally no longer economic to maintain the flock in production.

Broodiness

Broodiness is relatively uncommon in Pekin parent stock and egg layers, but can occasionally cause economic loss. In the trials reported in Chapter 6 (this volume), broodiness only occurred where birds were given a rapid increase in daylength when coming into lay, and experience in Europe shows that broodiness is uncommon when birds are held on a constant 17 h daylength.

Breeding Performance

Figures 6.42 and 7.10 describe the rate of lay that can be achieved by large commercial flocks of Pekin parent stock, but consistent performance of this nature depends upon providing reliable systems for the supply of feed, water, controlled daylength and air quality along with high standards of hygiene, management and husbandry from day-old to the end of lay.

Many factors affect laying performance, but experience shows that a relatively small number (see Table 7.3) is responsible for the majority of flocks showing inconsistent and reduced laying performance. Figures 7.11–7.13 describe the effect of the principal causes of poor laying performance on age at sexual maturity, rate of lay and laying performance to a fixed age.

Table 7.3. Principal factors affecting laying performance.

Event	Cause	Effect	Prevention
Variation in age at sexual maturity	Variation in: (a) 18 week restricted fed live weight	(a) See Fig. 6.25	Achieve 85% of mature weight at 18 weeks
	(b) Daylength during rearing 5–26 weeks of age	(b) See Figs 6.34 and 6.35	Provide a 17 h daylength during rearing and lay
	(c) Age at increase in daylength	(c) See Fig. 6.37	In cool climates provide 2 h feeding time at 18 weeks increasing to about 8 h by 22 weeks
	(d) Quantitative feed restriction when coming into lay	(d) See Figs 6.15 and 7.5	
Variation in pattern of lay	(a) Sharp increase in daylength at about 18 weeks	(a) See Fig. 6.36	Provide a 17 h daylength from day-old
	(b) Rigorous feed restriction in lay	(b) See Figs 6.15 and 7.12	In cool climates provide at least 6 h feeding time a day
Depressed rate of lay	Salpingitis caused by: (a) Rearing birds in proximity to other poultry		Rear in relative isolation from other poultry and maintain high standards of hygiene
	(b) Poor standards of hygiene during rearing	See Fig. 7.12	
	Nutrition (b) Mycotoxins	Reduced rate of lay and high mortality	See Chapter 5
Sharp fall in rate of lay	(a) Egg drop syndrome	See Fig. 7.13	Implement strict hygiene control, seek veterinary advice
	(b) Failure in water supply	See Fig. 7.13	Drinker failure: provide two drinkers in each pen Supply failure: provide standby water storage

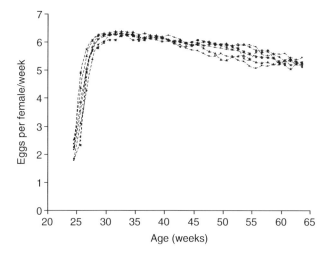

Fig. 7.10. Rate of lay of six large commercial flocks of Pekin parent stock hatched at about the same time of year and given similar environment, nutrition and husbandry during lay. Age at sexual maturity was 26 weeks, average egg production over 40 weeks based upon number housed was about 225 eggs per female, mature egg weight was about 89 g and mortality, infertility and hatchability were 7.65%, 4.8% and more than 80%, respectively. (Data from Cherry, 1993.)

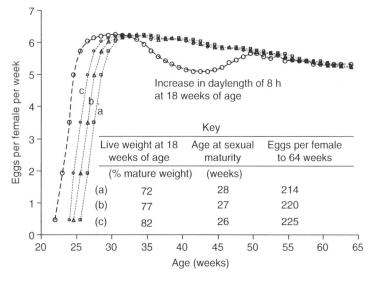

Fig. 7.11. The relationship between live weight at 18 weeks in the range of 72–82% of mature weight for birds given a constant daylength of 17 h on age at sexual maturity and laying performance to a fixed age, and the effect of an 8 h increase in daylength from 9 to 17 h at 18 weeks on age at sexual maturity and subsequent pattern of lay. (Data from Cherry, 1993.)

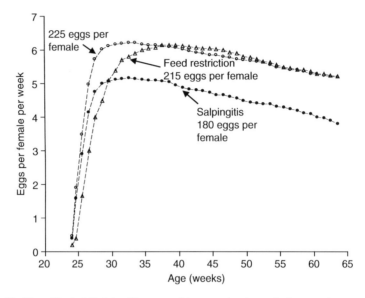

Fig. 7.12. The effect of Salpingitis caused by poor hygiene during rearing on subsequent laying performance, and of continued rigorous quantitative feed restriction to 28 weeks on age at sexual maturity and laying performance.

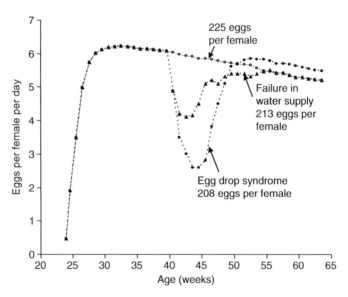

Fig. 7.13. Typical effects of 'egg drop syndrome' and short-term failure in water supply on subsequent pattern of lay and laying performance to a fixed age.

References

Cherry, P. (1993) Sexual maturity in the domestic duck. PhD thesis. University of Reading, Reading, UK.

Das, G.B., Haque, M.E., Ali, M.L., Chandra, G.C. and Das, C. (2003) Performance of Indian Runner, Zending and Khaki Campbells in integrated duck-cum-fish farming systems. *Pakistan Journal of Biological Sciences* 6(3), 198–201.

Dean, W.F. (1985) Nutrient requirements of meat type ducks. In: Farrell, D.J. and Stapleton, P. (eds) *Duck Production Science and World Practice*. University of New England, Armidale.

Elkins, R.G. (1987) A review of duck nutrition research. *World's Poultry Science Journal* 43, 84–106.

Hutt, F.B. (1952) The Jansen Khaki Campbell ducks. *Journal of Heredity* 43, 277–281.

Nho, L.T. and Tieu, H.V. (1996) Egg production and economic efficiency of Khaki Campbell ducks reared on locally available feedstuffs in the coastal land stretch of the Red River Delta. *Livestock Research for Rural Development* 9(1), 1–4.

Pan, C.M., Lin, E.I. and Chen, P.C. (1981) Studies on laying duck nutrition. 2. Protein and energy requirements of Tsaiya. *Journal of Taiwan Livestock Research* 14, 39–44.

Scott, M.L. and Dean, W.F. (1991) *Nutrition and Management of Ducks*. M.L. Scott of Ithaca, Ithaca, New York.

Shen, T.F. (1985) Nutrient requirements of egg-laying ducks. In: Farrell, D.J. and Stapleton, P. (eds) *Duck Production Science and World Practice*. University of New England, Armidale, New South Wales, Australia, pp. 16–30.

Valez, A., Brun, J.M. and Rouvier, R. (1996) Crossbreeding effects on reproductive traits in two strains of (Anas platyrhychos): Brown Tsaiya and Pekin. *British Poultry Science* 37(3), 571–577.

8 Fertility and Hatchability

Duck eggs take 28 days to hatch and are usually 'candled' (i.e. examined by shining a light through the eggs) on the tenth day to remove 'infertile' eggs. Fertility is reckoned as the percentage of eggs set which appear viable at 10 days, although this results in classifying very early embryo mortality as infertility. Eggs are normally transferred from setting to hatching machines between 22 and 24 days of incubation. Embryonic survival is usually calculated by expressing the number of chicks hatched as a percentage of the eggs transferred to the hatcher and this is called the hatchability of fertile eggs.

Effects of Selection

A commercial trial comparing selected and unselected genotypes confirmed that selection over about 20 years for efficiency of feed conversion and either growth rate or body composition substantially increased both rate of gain and mature bodyweight, but reduced laying performance by more than 30% and hatchability of fertile eggs to less than 78%. Conversely, pedigree selection for egg production and hatchability, using 100 single drake pens of a Pekin genotype (mature live weight about 2.2 kg), reduced age at sexual maturity, improved laying performance to more than 235 eggs per female in 40 weeks and increased fertility to about 97% and hatchability of fertile eggs to more than 90%. Clearly, reproductive performance can be improved by appropriate selection but, if selection is practised for meat traits alone, fertility and hatchability will decline.

Breeding companies attempt to offset the effects on breeding performance of selection for growth, efficiency and increased breast meat by mating males, selected solely for these characters, to cross-bred females derived from sires selected for meat traits to dams selected for laying and hatching performance. This approach, along with control of growth rate and daylength (see Chapter

6, this volume), high standards of husbandry and controlled egg weight (see Chapter 7, this volume), offers integrated companies the best method of optimizing breeding efficiency (see Figs 6.42 and 7.10) without substantially reducing the growth performance of the final cross-bred ducklings.

Fertility

Trials investigating the effect of genotype on fertility have confirmed that between 40% and 50% of eggs removed as infertile are dead germs, with most embryos dying at about 2–3 days incubation shortly after developing the blood streak. Analysis for the effect of age on 'infertility' showed that the percentage of dead germs in eggs removed as infertile increased in all genotypes between 26 and 34 weeks of age from about 3% to over 40%.

Other trials showed that substantially increasing the concentration of vitamins and trace elements in the feed supplied to parent stock had no effect on fertility or on the percentage of dead germs.

To achieve optimum fertility, it is important to rear males with females from day-old to 'imprint' the males (see Chapter 6, this volume). Trials investigating the effect of the ratio of ducks to drakes on fertility for selected Pekin genotypes (mature bodyweight 4–5 kg), with between 4 and 9 ducks per drake in large pens, showed that increasing the ratio beyond 5 ducks per drake reduced average fertility linearly by about 2.5% for every extra female placed. Similar trials with egg-laying genotypes (mature body weight 2–3 kg) confirmed that there was no economic advantage in reducing the ratio below about 7 ducks per drake. Increasing the ratio in equal steps to 10 ducks per drake reduced average fertility linearly by less than 1% for each addition of one duck to the female/male ratio.

Hatchability

The majority of embryos which fail to hatch die between 22 and 27 days of incubation. These are commonly termed 'dead-in-shell' and can be broadly divided into three categories. The first and most frequent type grow and develop normally, but appear to make no attempt to break through the shell. They consequently die at about 28 days. Embryos of the second type die at the same age, but show a characteristic flattening and bending of the beak along with oedema and haemorrhage in the hatching muscle behind the head. These are symptoms of sustained but unsuccessful efforts to break out through the shell. The incidence of this condition declines with increasing breeder age. The third type die between 22 and 28 days because they become malpositioned during their development in a way that prevents them from breaking out of the shell and hatching.

Shell Structure and Weight Loss

Egg shells are porous to allow gas exchange during incubation. Balance between consumption of oxygen and production of carbon dioxide is achieved through loss of water (Hoyt *et al.*, 1979). Rate of loss is proportional to 0.75 power of egg weight, and about 16% of fresh egg weight is lost during incubation to 28 days (Drent, 1975). This is considered optimal to achieve good hatchability. However, the rate of moisture loss increases substantially when chicks breach the shell prior to hatching and so a weight loss of about 12% to 24 days (or an average of 0.5% per day) is probably optimal for good hatchability.

Unreported trials measuring weight loss from individual eggs (sample size 1000 eggs) confirm that weight loss during incubation to 24 days follows a normal distribution with a coefficient of variation of about 20%. Thus, embryos can tolerate a wide range of weight loss, but eggs which lose less than about 7% suffer poor hatchability. Some hatcheries routinely weigh samples of hatching eggs and adjust water vapour pressure in order to control mean weight loss during incubation.

In trials investigating the effect of genotype and age on weight loss during incubation of fertile eggs from selected and unselected Pekin breeding flocks aged 27 weeks, the average weight loss to 25 days from eggs (average weight 82.6 and 79.2 g) was 0.52 and 0.55 g per day from eggs that hatched but 0.47 and 0.49 g per day from dead-in-shell. Individual eggs in this and subsequent trials showed an increased weight loss of about 0.005 g per day for each 1 g increase in egg weight in the range of 70–100 g, but there was no effect of egg weight on weight loss measured as a percentage of the starting weight.

Two consecutive trials which measured weight loss of eggs (average weight about 90 g) from individual females in single drake pens showed that weight loss for eggs from about 175 females providing more than five fertile eggs in both trials was repeatable ($r^2 = 0.69$; see Fig. 8.1). Mean daily weight loss to 21 days in both trials was about 0.49% with a coefficient of variation of 22%, and variation in daily average percentage weight loss to 21 days of eggs from individual females was about 16%. There was no substantial difference in these trials between the weight loss to 21 days of eggs that subsequently hatched (0.53% per day) and fertile eggs that failed to hatch (0.47% per day). However, in both trials less than 25% of females provided more than 55% of dead-in-shell, with some females contributing a substantial percentage of dead-in-shell in both trials (see Fig. 8.2). A similar analysis showed that a large percentage of eggs from a small number of females were infected in both trials, suggesting that shell porosity and nest hygiene are not the only factors influencing infection of hatching eggs.

Although hatchability shows reasonable repeatability for individual ducks (see Fig. 8.2) caution should be exercised before concluding that this indicates a heritable component. Ducks carrying infection in their ovaries will repeatedly yield a high proportion of dead-in-shell and it is far from clear that infection (or susceptibility to infection) is a heritable trait. Nevertheless, the results of pedigree selection, as reported above, show that hatchability (and fertility) can be improved by selecting sires whose daughters give the best breeding performance.

Measuring buoyancy of hatching eggs also provides a simple and low-cost method for measuring weight loss (see Fig. 8.3). Convention suggests that weight loss is satisfactory if about 90% of the eggs sampled float when placed in warm water at about 24 days. Trials investigating the relationship between

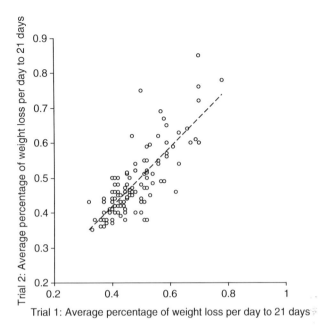

Fig. 8.1. The relationship between average weight loss of eggs during incubation from females providing more than five fertile eggs in two consecutive trials. Eggs were identified to individual female and egg weight was recorded prior to incubation and at 21 days.

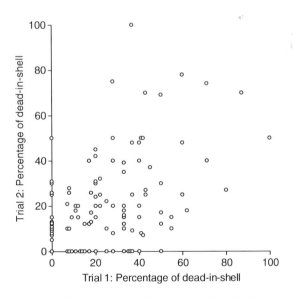

Fig. 8.2. The relationship between percentage of dead-in-shell recorded from breeding females providing more than five fertile eggs in consecutive trials.

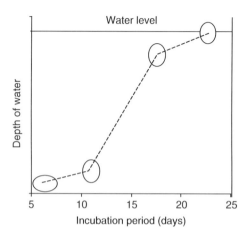

Fig. 8.3. Effect of incubation for about 6, 12, 18 and 24 days on buoyancy of fertile hatching eggs placed in warm water.

weight loss and buoyancy confirmed that 89% of fertile eggs (sample size 949 eggs) floated when placed in a bath of warm water after 23 days incubation. The 'floaters' had an average weight loss of 11.5% and a hatchability of 91%, while the 'sinkers' had an average weight loss of 7.9% and a hatchability of 82%.

Ducks lay eggs coated with a waxy cuticle to provide a barrier against infection. However, wildfowl use their webbed feet to turn eggs during incubation and the frequent abrasion reduces cuticle thickness and probably encourages gas exchange to increase in a curvilinear manner during incubation.

In China and the Far East, traditional methods are still used to incubate duck eggs. Artificial heat is provided for the first 5 days of incubation when eggs are placed several layers deep on beds and covered with down-filled quilts. Developing embryos produce sufficient heat to maintain temperature at about 37°C. Eggs are turned by hand several times daily and this erodes the cuticle, increasing porosity and gas exchange and probably improving subsequent hatchability. These traditional hatcheries regularly achieve more than 75% hatchability of all eggs set.

Many commercial companies spray hatching eggs daily during incubation, no doubt to simulate the damp environment experienced during incubation in the wild. However, trials in a temperate climate suggest that spraying eggs in this manner has no appreciable effect on hatchability. On the other hand, the evidence presented in Fig. 8.4 shows that washing eggs from mature breeding flocks in a detergent solution increased weight loss during incubation and suggests that spraying eggs daily erodes the cuticle, increasing porosity and possibly hatchability in hot and humid environments.

Figure 8.4 shows that removing the waxy cuticle from hatching eggs of an unselected Pekin genotype with a hypochlorite solution increased weight loss to more than 11.2% in 24 days compared to about 10% for unwashed eggs. Weight loss of eggs washed in a sanitant and detergent solution in this and other unreported trials increased with parent flock age, suggesting that cuticle thickness declines as parent stock get older.

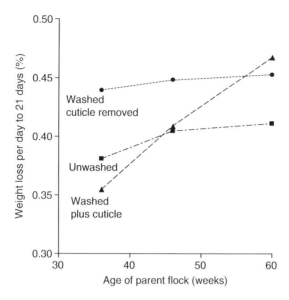

Fig. 8.4. Effect on weight loss to 21 days of washing hatching eggs from Pekin parent stock in either a hypochlorite solution, which removes the cuticle, or an egg sanitant, which does not.

Controlled Growth and Daylength

Manipulating feed and light programmes can have large effects on growth and sexual maturity, and these in turn affect fertility and hatchability, especially in the early stages of lay.

Figure 8.5 shows the effect on early infertility of controlling growth and age at increase in daylength for birds given step-up lighting programmes. Birds in Trial 8 were grown to achieve about 70% of mature *ad libitum* fed live weight at 26 weeks and were maintained at this weight by continued feed restriction until it became apparent that restricting live weight in this manner was affecting performance. Feed allowance was then increased from about 30 weeks, and live weight increased to about 76% of mature (*ad libitum* fed) live weight by 34 weeks. In Trial 9, daylength was stepped up from either 14 or 24 weeks and the birds were grown to achieve 3.6 kg (about 75% of mature *ad libitum* fed live weight) by 26 weeks and then given 5 h feeding time from about 28 weeks. Early infertility reflected the age at which each group came into lay. In Trial 8, the flocks did not reach 90% fertility until 35 weeks, whereas in Trial 9, with more generous feeding, this level was reached at 32 weeks. Before 32 weeks, the birds in Trial 9 that had been stimulated from 14 weeks had better fertility than the later-maturing group that was left on short days until 24 weeks.

The more generous feeding in Trial 9 also improved early hatchability (see Fig. 8.6), but the different ages of light stimulation had no clear effect on hatchability. However, yields to 55 weeks were 143 eggs per bird for the group

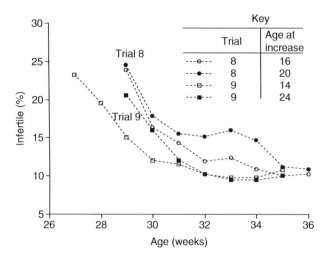

Fig. 8.5. Effect of age at increase in daylength upon fertility for birds reared on step-down, step-up lighting programmes. Birds of the same genotype in two trials were reared on 23h light to 8 weeks and then given 8h to 16 or 20 weeks in trial 8 and 14 or 24 weeks in trial 9, after which daylength was increased by 1h per week to 17h. (Data from Cherry, 1993.)

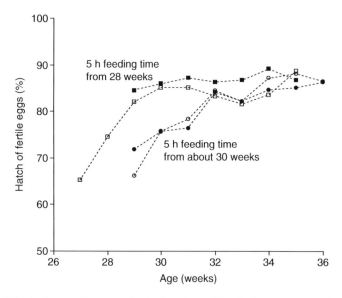

Fig. 8.6. Effect of age at increase in daylength and feed allowance upon hatchability of fertile eggs in two trials for birds reared on step-down, step-up lighting programmes. Treatments and key as in Fig. 8.5.

stimulated from 14 weeks compared to 129 eggs per bird for those stimulated from 24 weeks.

Step-up lighting programmes are widely used to stimulate egg production in both laying and breeding ducks. Using artificial light to increase daylength from the prevailing natural level to about 17 h, either rapidly or in a single step, is widely believed to encourage male fertility. However, Figs 8.7 and 8.8 show that birds given either a gradual or a rapid increase in daylength had similar early fertility and hatchability. Overall performance was similar for the two treatments (142 and 139 eggs per bird to 55 weeks, 86% and 84% fertility and 85% and 83% hatchability for gradual and rapid increases, respectively).

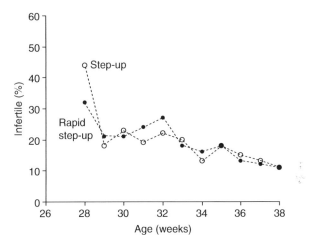

Fig. 8.7. Effect of rearing birds on step-up and rapid step-up light programmes on fertility. Birds were reared in lightproof accommodation and given 8 h light from 8 to 18 weeks when daylength was increased by either 1 h per week to reach 17 h at 26 weeks or by 4 h, and then by equal increments to provide 17 h by 21 weeks of age. Growth was controlled to achieve about 75% of *ad libitum* fed mature live weight at 18 weeks of age. (Data from Cherry, 1993.)

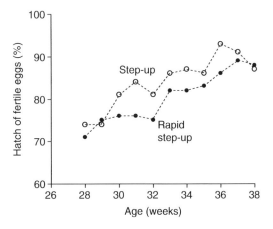

Fig. 8.8. Effect of rearing birds on step-up and rapid step-up light programmes on hatchability of fertile eggs. Accommodation, lighting treatments and controlled growth as in Fig. 8.7.

Increasing natural or artificial daylength by more than 3 h during the later stages of rearing can reduce post-peak rate of lay and laying performance to a fixed age compared with birds reared on a constant daylength of 17 h or a step-down lighting programme (see Figs 6.36, 6.39 and 6.42). In Europe, breeding flocks reared on natural daylength during the winter months and then given a single step-up to about 17 h at about 20 weeks of age to stimulate egg production in early spring suffered a photorefractory response, which substantially reduced post-peak rate of lay, fertility and hatchability and breeding performance to a fixed age.

Figure 8.9 describes the effect of providing step-down and step-up lighting programmes, while controlling growth, on early fertility of Pekin genotypes A and B. Step-down lighting and restricting growth up to 26 weeks delayed sexual maturity and increased early infertility, particularly in genotype B. However, Figs 8.9 and 8.10 show that neither lighting programme nor the different growth patterns had any substantial effect on fertility or hatchability beyond about 30 weeks of age.

Figures 8.11 and 8.12 describe the effect of controlled growth and either constant or step-down/step-up lighting on fertility and hatchability. The data

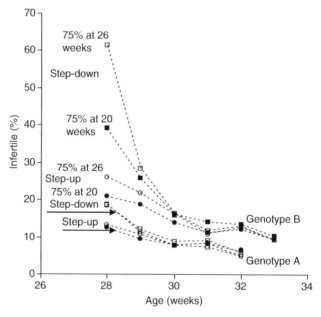

Fig. 8.9. The effect on fertility of rearing genotypes A and B on either step-down or step-up lighting programmes and on two controlled growth programmes. Genotypes A and B with mature *ad libitum* fed live weights of about 4 and 5 kg, respectively, were reared to achieve 75% of mature live weight at either 20 or 26 weeks of age. All birds were given 23 h daylength to 8 weeks, and then either a step-down light programme to 17 h at 18 weeks or a step-up programme where daylength was reduced to 8 h at 8 weeks and then increased weekly from 18 weeks to reach 17 h at 26 weeks of age. (Data from Cherry, 1993.)

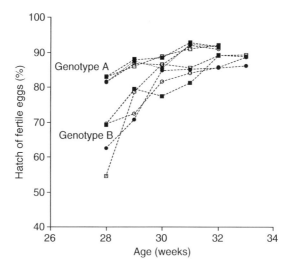

Fig. 8.10. Effect of genotypes (A and B), step-down or step-up lighting programme and controlled growth to 20 or 26 weeks on subsequent hatchability of fertile eggs. Treatments and key as in Fig. 8.9.

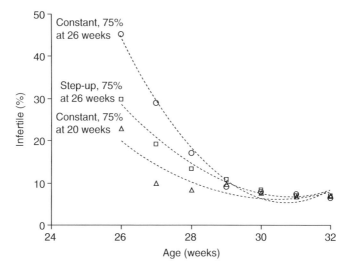

Fig. 8.11. Effect of rearing commercial flocks on either a step-down/step-up or a constant daylength on fertility. Birds were reared on either (i) 23h light from day-old to 5 weeks followed by natural daylength to 18 weeks when daylength was increased in equal steps using artificial light to provide 17h daylength at 26 weeks; or (ii) were given a constant 17h daylength from day-old. The birds given constant daylength were reared to achieve about 75% of mature *ad libitum* fed live weight at either 20 or 26 weeks of age and given similar controlled feeding time in lay. (Data from Cherry, 1993.)

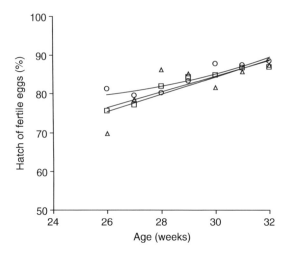

Fig. 8.12. Effect of rearing commercial flocks on either step-down/step-up or constant daylength on hatchability of fertile eggs. Lighting treatments, controlled growth and key as in Fig. 8.11.

confirm that feeding to achieve about 80% of *ad libitum* fed mature live weight by 20 weeks substantially improved fertility up to 32 weeks for flocks reared on a constant 17 h daylength. The results of these trials, further evidence presented in Chapter 6 (this volume) and commercial experience (see Fig. 6.42) all confirm that controlling growth to achieve about 80% of mature *ad libitum* fed live weight by about 18 weeks will optimize breeding performance to a fixed age for birds reared and maintained in lay on a constant daylength.

Lighting for Males

Figure 8.13 shows the effect on early fertility for drakes given one of eight rearing programmes. They were mated to ducks which were reared on a constant 17 h daylength and reached about 80% of mature live weight at about 20 weeks. The treatments applied to the males had no significant effect on early fertility. This is presumably because drakes reach sexual maturity sooner than ducks, so that males from all eight treatments were fully fertile before they were introduced to the duck pens at 22 weeks.

Commercial experience confirms that placing mature drakes with females prior to sexual maturity does not improve early fertility.

Season and Hatchability

Commercial experience with Pekin genotypes both in Europe and in North Carolina confirms that hatchability declines during the summer months. Analysis for the effect of season on hatchability of duck eggs over the years

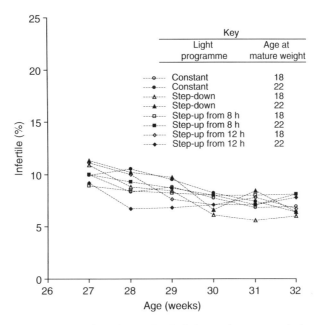

Fig. 8.13. Effect on fertility of rearing males in lightproof accommodation with various light programmes and controlled growth. Males were reared to achieve about 80% of their potential mature *ad libitum* fed live weight of about 5 kg at either 18 or 22 weeks of age and given one of the following lighting programmes:
1. Constant daylength of 17 h;
2. Step-down, 23 h to 6 weeks when daylength was reduced in equal increments to provide 17 h at 18 weeks;
3. Step-down, step-up, 17 h to 6 weeks when daylength was reduced to 12 h and then increased weekly in equal increments to provide 17 h by 18 weeks;
4. Step-down, step-up, 17 h to 6 weeks when daylength was reduced to 8 h and then increased in equal increments weekly to provide 17 h at 18 weeks.
Females were given a constant 17 h daylength and reared to achieve about 80% of mature weight by about 20 weeks of age.
Trial design: four light programmes by two controlled growth treatments provided eight treatments with two replicate pens making a total of 16 pens. Twenty males were placed with 100 females in all pens at 18 weeks and given timed feeding from either 18 or 22 weeks when birds were given 2 h feeding time increasing by an hour a week to 6 h. (Data from Cherry, 1993.)

1995–2002 recorded at Khulna University in Bangladesh (Chowdhury *et al.*, 2004) shows that hatchability averaged 57.7% in winter (November–February), 54.1% in summer (March–June) and 49.6% during the monsoon (July–October). This shows that hatchability in a tropical monsoon climate is affected by season, even for a fully acclimatized local genotype.

A trial investigating the effect of environmental temperature on breeding performance of a selected Pekin genotype, using artificial heat to raise house temperature to about 27°C, found reduced egg shell porosity with weight loss to 24 days of incubation below 10%. Fertility was reduced by 2% and hatchability of fertile eggs by 7% compared with birds maintained at less than 17°C. Subsequent trials found that reducing water vapour pressure, so increasing weight loss during incubation, did not improve hatchability, but removing cuticle by washing eggs in a hypochlorite solution and then adjusting water vapour pressure to control weight loss during incubation substantially improved hatchability, in both the cool and the hot environments.

High temperatures combined with high humidity in the Far East, and during the summer in some regions of Europe and North and South America mean that optimum weight loss can sometimes only be achieved if water vapour pressure is artificially reduced. However, this is not cost-effective in commercial hatcheries. Some loss in hatchability appears to be an inevitable consequence of maintaining highly selected Pekin genotypes in hot and humid climates.

Removing cuticle increases porosity, but hatching eggs no longer enjoy the protection provided by cuticle against infection. Commercial experience confirms that high standards of hygiene are essential to prevent infection of these eggs during incubation.

In hot and humid climates, eggs stored at about 15°C frequently 'sweat' when removed from storage prior to incubation, encouraging subsequent infection. Condensation can be prevented either by increasing environmental temperature gradually or by providing sufficient air movement to prevent the temperature of air around the eggs from falling below the 'dew point'.

Other factors can also profoundly affect hatchability from parent stock maintained at high temperature. For example, increasing ambient temperature in late spring and early summer is often associated with a decline in post-peak breeding performance but, as noted above, this can also be a photorefractory response if parent stock are given a sudden increase in daylength in early spring to stimulate sexual maturity. Similarly, in the tropics, stepping up daylength to 17 h at about 18 weeks for parent stock reared on a natural day of about 12 h can reduce post-peak breeding performance.

As we have seen, controlling growth substantially improves hatchability. Breeding companies in Europe and North America normally provide recommendations for feed restriction of their parent stock, based on experience gained in either temperate or continental climate. However, evidence from a trial in Singapore (see Fig. 6.11) indicates that very high temperatures during rearing will affect food intake, growth and subsequent mature live weight. To delay sexual maturity and achieve optimum breeding performance (see Table 6.4 and Fig. 6.49) under these conditions it is important to relate restriction to the *ad libitum* growth to be expected at the prevailing temperature.

Data presented in Chapter 7 (this volume) confirm that ambient temperatures above about 25°C reduce both feed and nutrient intake and subsequent rate of lay. Many breeding companies increase nutrient concentration during summer in the hope of improving breeding performance. However, unreported trials at high ambient temperature during summer in the USA and in various European countries found that increasing nutrient concentration by about 20% for both restricted and *ad libitum* fed birds had no substantial effect on fertility or hatchability. There appears to be no economic advantage in increasing nutrient concentration beyond that described in Table 7.2.

Egg Quality

Egg quality is affected by genetic, maternal and environmental effects. The most obvious genetic effect is shell colour; different breeds produce white, green and blue egg shells.

Empirical experience and trials with different genotypes confirm that selecting eggs on the basis of their shape, texture and appearance substantially improves hatchability, but hatcheries, when short of hatching eggs, often incubate substandard eggs and accept the lower hatchability to achieve their production requirements. Eggs from genotypes selected for breeding performance are more uniform and have better shell quality and consequent hatchability than eggs from genotypes selected for growth and meat characters. Pedigree selection for egg quality and breeding performance are essential to prevent serious decline in hatchability when selecting for improved growing performance.

Common defects include cracks, damaged shells and abnormal shapes and shell textures. About 2% of eggs are commonly removed as cracked on farm, but often more than 5% of eggs incubated may have fine hairline cracks which affect hatchability. Trials at different locations confirm that eggs with hairline cracks experience excessive weight loss (>18% to 24 days), reducing their hatchability to less than 50%. Hairline cracks also encourage infection. Infected eggs produce purulent material, which infects both adjacent eggs and those on lower trays in the incubator, substantially reducing hatchability and affecting air quality and chick health.

Double-yolked eggs never hatch and small eggs (<50 g) do not hatch well. Controlling growth and achieving sexual maturity at about 26 weeks of age for selected Pekin reduces but does not altogether prevent the incidence of double-yolked and small eggs. However, the proportion of these declines at a rate of about 1.5% per week from about 12% at 25 weeks to less than 2% by 35 weeks. The incidence of rough and porous, thin shells and hairline cracks usually increases with age and can also be affected by climate when a high effective temperature reduces nutrient intake.

Husbandry

Space allowance and litter and nest husbandry will all affect egg hygiene. Wood shavings make a suitable nesting material. While straw is acceptable as floor litter, it is not suitable for nests as it soon becomes soiled and wet.

Placing more than about two birds per m² of floor space for parent stock housed intensively on litter or 2.5 birds per m² on litter with slatted flooring reduces laying performance and adversely affects egg hygiene and subsequent hatchability.

In hot climates, providing artificial light early in the morning encourages early laying and provides an opportunity to collect, wash and store eggs at a controlled temperature when it is cool, preventing further cell division in the germinal disc from adversely affecting hatchability. In warm and temperate climates limiting feed intake (see Figs 7.5–7.7) prevents egg weight for selected Pekin genotypes from exceeding about 92 g and adversely affecting hatchability.

One nest should be provided for every four laying ducks housed. Nests should be located along sidewalls and pen partitions to encourage birds to lay their eggs in clean well-littered nests. Floor eggs, even when apparently clean, are invariably heavily contaminated with microorganisms.

Protecting nests with portable guards prevents birds from soiling nests during the day. Placing a small amount of fresh litter in nests before removing the guards in the late afternoon will substantially reduce the incidence of soiled eggs the following day.

Collecting eggs onto stackable plastic trays (turkey egg trays are widely used) prevents egg to egg contact and reduces the incidence of hairline cracks. It is important to wash hatching eggs in a suitable sanitant as soon as possible after collection and to store them at a controlled temperature and humidity (12–15°C and about 75% relative humidity). Dirty eggs and floor eggs should be identified with a waterproof mark, washed and then incubated separately from nest eggs because they suffer substantially higher levels of infection, even after washing, and will adversely affect the hatchability of clean eggs.

After eggs are delivered to the hatchery, it is advisable to allow them to rest for about 24 h before incubation. Storing eggs in incubator trays and trolleys facilitates turning during storage. It is advisable to turn stored eggs several times a day and important to avoid storing eggs beyond about 7 days to prevent loss of hatchability.

Improving Hatchability

Genetic selection for growth, body composition and efficiency of feed conversion adversely affects hatchability (see Fig. 8.14). However, controlling growth and daylength during rearing to delay sexual maturity, along with feed restriction to control egg weight and cuticle removal to improve gas exchange during incubation can all help to improve hatchability, with the net result that there has been no substantial decline in hatchability over the past decade.

High standards of husbandry and regular supervisory pen checks along with statistical process control (see Chapter 4, this volume) substantially reduce the incidence of cracked eggs, improve egg hygiene and reduce subsequent infection during incubation, but have only a limited effect on either fertility or incidence of dead-in-shell, which is responsible for about 75% of overall hatching loss (see Fig. 8.15).

Hatchability is a discontinuous variable; an egg either hatches or it does not. Studies on hatchability can only be carried out on groups of eggs and very large egg numbers are required to detect small improvements in hatchability. About 10,000 eggs per treatment are required if differences of 1% in hatchability are to be detected (Laughlin and Lundy, 1976). Replicate samples can be used to achieve sufficient numbers to satisfy statistical requirements, but only where factors such as genotype and environmental conditions remain unchanged. However, few production companies have the facilities required to

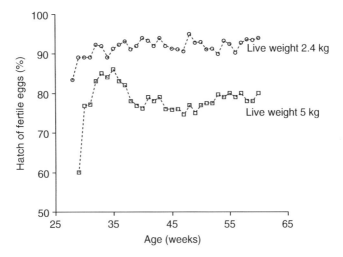

Fig. 8.14. Hatchability of fertile eggs for Pekin elite genotypes selected for breeding performance and growth, efficiency of feed conversion and body composition with mature *ad libitum* fed live weights of about 2.4 and over 5 kg, reared on feed restriction to 26 weeks of age and given 17 h daylength in lay.

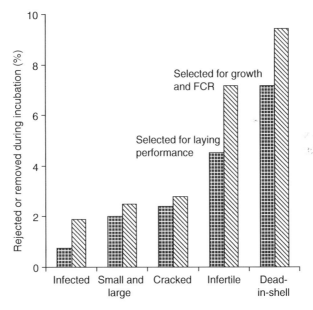

Fig. 8.15. Factors affecting hatchability of genotypes selected for laying performance or for table duckling production with mature *ad libitum* fed live weights of about 2.4 kg and more than 4 kg, respectively. The data recorded over 4 consecutive weeks describe the average percentage rejected before or during incubation from more than 600,000 eggs provided by parent flocks (average flock size about 2000 between 30 and 68 weeks of age) reared intensively with feed restriction and controlled daylength during lay in a cool temperate climate. Hatchability of all eggs incubated for genotypes was 87.4% for the line selected for egg production and 82.3% for the line selected for growth.

carry out trials on this scale and this may explain why hatchability and fertility have not improved over the last 20 years.

An alternative approach, proposed by Laughlin and Lundy (1976), entails measuring the effect of treatments on continuous variates affecting physical and biological characteristics and subsequent hatchability. Experience confirms that using this approach to investigate the effect of shell porosity on gas exchange during incubation helped to reduce embryo mortality and substantially improved commercial hatchability.

References

Cherry, P. (1993) Sexual maturity in the domestic duck. PhD thesis, University of Reading, Reading, UK.

Chowdhury, M.M.I., Ashraf, A., Mondal, N.M.A.A.M. and Hasan, M.M. (2004) Effect of season on hatchability of duck eggs. *International Journal of Poultry Science* 3(6), 419–421.

Drent, R.H. (1975) Incubation. In: Farmer, D.S. and King, J.R. (eds) *Avian Biology*, Vol. 5. Academic Press, New York, pp. 351.

Hoyt, D.F., Board, R.G., Rahn, H. and Pagnelli, C.V. (1979) The eggs of Anatidae: Conductance, pore structure and metabolism. *Physiological Zoology* 52, 438–450.

Laughlin, K.F. and Lundy, H. (1976) The influence of sample size on the choice of method and interpretation of incubation experiments. *British Poultry Science* 17, 53–57.

9 Genetic Improvement

The remarkable improvements in growth rate and egg production that were achieved in domestic ducks during the 19th and 20th centuries have been described in Chapter 1 (this volume; see Fig. 1.1). From an original Mallard genotype that grew to about 1200 g over a period of 3 months and laid one or two clutches of 12–15 eggs in a year, breeders have, through many generations of selection, derived lines capable of reaching 3 kg live weight in 5 weeks and other lines that lay more than 300 eggs in a year. However, no one has produced a line that both lays 300 eggs *and* grows to market weight in 5 weeks and no one should expect to do so. In ducks, as in all other farm animals, there are strong negative genetic correlations between meat characteristics and reproductive traits. This means that any programme of genetic improvement needs to specify clearly what objectives are sought and, of course, those objectives must be related to the market that is to be supplied.

Objectives for Egg-producing Ducks

Production of duck eggs for the table is not important in Europe or the Americas because, in those places, chicken eggs are cheaper to produce and they meet the needs of the consumer very well. But in India and Asia vast areas of wetlands and open water are used to keep laying ducks; here duck eggs are an important part of the local diet.

Laying flocks are mostly capable of producing around 300 eggs per duck and where this figure is not achieved it probably has more to do with seasonal shortage of food than lack of genetic potential (Thummabood, 1992). If, in a particular area, the local breed is less productive even when well fed, indicating a need for genetic improvement, the obvious route to take is breed substitution.

Given a stock of ducks with a good rate of lay, the main concern of the breeder may well be to improve secondary traits such as egg size, shell strength or hatchability.

Methods of Selection for Egg Production

Within-line selection of females for egg production is expensive, because it requires a recording system, and is slow, because the heritability of egg number is low (0.1–0.3 in most populations (Cheng *et al.*, 1995, 2002; Chapuis and Larzul, 2006). Efficient genetic improvement of egg numbers can only be made by progeny testing or sib testing males and, since this requires identification of sire pedigree at hatching, it is a costly business. Much faster and cheaper progress will be made by buying in a strain of ducks with good laying potential either to substitute for the local breed or to cross with it.

The choice between breed substitution and upgrading local stocks by cross-breeding will depend upon the availability of capital, the difference in performance characteristics between the indigenous and introduced strains and the target genotype that is sought. In many cases, the purchase of a completely new stock of males and females would be prohibitively expensive for duck keepers living at subsistence level, whereas the purchase of enough day-old drakes to cover all the ducks in a flock might be feasible. Subsequent inter-breeding of the first crosses can be expected to yield a stock with egg production about halfway between the two founding genotypes.

When selection within egg-producing strains is practised, it is more often for characters such as egg quality or hatchability than for egg number. Shell strength and appearance may not be problems in the first 200 eggs laid but, later on, these characters usually deteriorate, so affecting the proportion of eggs fit to market or fit to incubate. Simple selection of females by independent culling is usually enough to maintain satisfactory egg quality, but use of an index incorporating all traits of importance in the particular strain is more efficient. For example, Cheng *et al.* (2003) describe the successful use of an index chosen to improve shell strength and egg number while holding body weight and egg weight constant.

Fertility and hatchability are usually better in egg-laying strains than in meat-producing ducks and there may be no need for further selection. If either or both of these traits do need genetic improvement, progeny testing is the only effective method, since both traits have low heritability (usually in the range 0.1–0.2). It should be standard hatchery practice to reject eggs that are abnormal in appearance, but this culling of extremes rejects only a small proportion of hatching eggs. It therefore exerts only a very small selection pressure and is more effective in preventing a negative drift than in achieving any real genetic improvement in hatchability.

Objectives for Meat-producing Ducks

Growth rate is often stated as the most important trait for selection in meat animals, but the reason for wanting faster growth is that it goes hand in hand with greater efficiency. Where the desirable weight at slaughter is relatively fixed, as with ducklings and broilers, the sooner the target weight is achieved, the lower will be the feed cost of maintaining the animal up to the point of slaughter. Housing and labour costs are also reduced by early finishing, although these factors have a smaller impact on overall profitability than the feed bill.

However, reaching market weight at an earlier age cannot be a sole objective. The composition of the carcass is important. A bird that is all skin and bone will not do; nor will one so immature that its bones have not adequately calcified. Duck down is also an important contributor to income in many duck enterprises. Selection for 'growth' therefore has to include breast meat development, calcification of the keel and feather development, as well as weight at a given age.

Another factor which becomes important when a line has been selected for rapid growth over a number of generations is mobility. Dyschondroplasia is a common problem in fast-growing birds and difficulty in walking can lead to serious losses. Mobility is particularly important in operations where ducks are driven onto trailers for transport to the processing plant, but this practice also provides a method of indirect selection for fitness.

Number of offspring per breeding female is also an important component of profitability in any meat enterprise. How this trait may be maximized, without unduly sacrificing meat characteristics, is described below.

Methods of Selection for Meat Production

The problem that soon arises when any type of animal is selected intensively for meat production is that reproductive potential declines. This is true of beef, sheep, pigs, chickens and turkeys. Ducks are no exception. The physiological reasons for these adverse genetic correlations are not clear and may not be the same in all species, but the genetic correlations seem to be inescapable.

Because it is not possible to breed a strain of ducks which has outstanding meat qualities combined with excellent egg production and hatchability, all large-scale companies producing ducks for meat use distinct parent lines for the final market cross. This means, of course, that the cross-bred ducklings grown for meat cannot themselves be retained as future breeders. Separate sire and dam lines must be maintained and the male duckling from the dam lines and female duckling from the sire lines have to be disposed of. Growing and selling these surplus chicks is a relatively unprofitable trade but is made worthwhile by the value that attaches to their siblings retained as parent stock.

The growth and efficiency of first-cross market duckling will not be as good as that of the pure-bred sire line but, because the dam line will produce perhaps twice as many day-old duckling, even when mated to sire-line males, the overall operation is much more profitable than a system using either of the pure parent lines.

Selection of Sire Lines

Growth rate (or weight at a fixed age) has a heritability in the range 0.4–0.6 (Xu *et al.*, 2004; Chapuis and Larzul, 2006) and therefore individual selection on the basis of phenotype is an efficient method of making progress.

Although selection for growth rate gives a substantial correlated improvement in feed conversion efficiency, direct selection for efficiency leads to faster improvement in this trait. This is because some of the genetic variation in food utilization efficiency is associated with factors other than earlier achievement of a given live weight. These factors include a shift in carcass composition towards a leaner bird (fat deposition is energetically costly) and a reduction in activity.

The commonest method of testing for feed efficiency has been to place birds in individual pens at 14 days of age and to record their feed consumption up to market weight. An alternative, which reduces errors caused by food wastage, is to supply a fixed quantity of feed to each bird each day and to measure final weight. In this case the data need to be adjusted for starting weight, since being small at the start of recording conveys an automatic advantage when the supply of feed is fixed. Another way of recording feed intake, which does not involve individual penning, is to fit a transponder to the wing of each duck and to combine this with feeders that record the identity of the bird and the amount of feed eaten at each visit.

If direct selection for efficiency is practised, this needs to be combined with some form of carcass assessment, because subcutaneous fat deposition will decline in response to selection for feed efficiency. This has implications for eating quality.

Breast meat percentage is positively correlated with live weight at a fixed selection age (Xu *et al.*, 2004) and so progress in this important trait can be made by selecting for growth alone, although faster improvement can be expected by using a selection index that includes breast meat as one of a number of traits. Clayton and Powell (1979) reported the heritability of breast muscle as a proportion of the eviscerated carcass to be 0.55. However, the genetic correlation of live weight with the combined yield percentage of breast and leg meat is virtually zero (Thiele, 1995), indicating that selection for faster growth (weight for age) shifts the proportions of body parts, without changing the proportion of muscle in the body as a whole. In other words, ducklings develop their legs first and their flight muscles (and feathers) later (see Fig. 5.38). This developmental process is speeded up by selection, leading to a bird which has invested more in its flight muscles by the time of slaughter.

Various devices have been developed to measure breast fat cover and muscle depth, and ultra sound probes are widely used. However, these methods are slow and do not simultaneously judge breast shape, keel shape or calcification. An experienced handler can make a multidimensional judgement of all these characteristics in less time than it takes to make a single objective measurement, and so subjective assessment may be preferable in practice. For academic studies, however, objective measures are required. For example, Xu *et al.* (2004) report a selection index constructed to improve breast meat weight by combining measurements of live weight, keel length, breast width and breast meat thickness.

Some companies use full carcass dissection of smaller samples of pedigree birds as a means of maintaining carcass quality in their most important sire lines.

Selection for fitness to walk is usually achieved by a combination of visual scoring of live birds and examination of an oblique section of the hock joint at the time of carcass analysis. These data are then combined with other criteria in an index to be used for the selection of sires on the basis of progeny performance.

Feather development can be scored visually at the same time that breast quality is assessed. Feather yield is an important economic trait in its own right, but early feathering is also correlated with early development of flight muscles and so selection for good feathering will aid selection for proportion of breast muscle at slaughter age (and vice versa). There is also good evidence that selection for the length of primary feathers at 35–40 days tends to reduce age at sexual maturity in females. This is useful because it offsets the extent to which selection for early growth rate increases mature body size.

Mature body size will increase in response to selection for early growth rate and this has implications for the cost of feeding the parent flocks. Some of this adverse effect can be offset by selection for length of the first primary feathers; however, in practice, the main method of controlling adult body size is by controlled feeding to prevent birds from reaching their full potential weight (see Chapter 6, this volume).

As sire lines are improved, their reproductive potential declines, particularly among the females. The breeder should resist the temptation to select for egg production and hatchability in his sire lines, unless it becomes very poor. That will necessarily lower the rate of improvement in growth and meat characteristics and so make the final cross less competitive. Although, with low productivity, more sire-line dams are needed to produce the required number of male parents, the number of grandparents needed is still relatively tiny. For example, to produce 2 million ducklings a year takes about 12,500 dams, assuming 160 chicks per parent female placed. These 12,500 dams must be mated to 2500 sires, but those can be produced from only 32 sire-line females, even if productivity in the sire line is as low as 80 (as hatched) ducklings per female per year. Each duckling placed bears only a tiny fraction (1/62,500) of the cost of maintaining its paternal grandmother. If, after many generations of decline, it does become necessary to select for chick production in a sire line, either because of low egg number or poor hatchability, that will require progeny testing of potential sires.

Egg size has a negative genetic correlation with egg number (Sochoka and Wezyk, 1971) but increases as a correlated response to increased (potential) adult body size. Again, egg size can be readily limited by controlled feeding in the growing stage followed, if necessary, by controlled feeding in the laying house (see Fig. 7.5).

Selection of Dam Lines

Dam lines are necessarily a compromise between meat- and egg-producing characteristics. New lines may be formed by crossing an egg-laying strain with an outstanding meat type.

Selection within an existing dam line may be based initially on individual selection of males and females for weight and conformation at market age, retaining at least twice as many birds at this stage as are needed to reproduce the line. Females can then be penned in sire groups with their pen performance recorded, allowing selection of retained males on the basis of their sisters' egg production and hatchability.

Precisely what selection is practised in dam lines will depend upon an assessment of the strengths and weaknesses of the particular line. It is not unusual to find that no selection is being practised, because the cost of pedigree hatching and recording is not justified by the small gains to be expected from selecting simultaneously for meat and reproductive characteristics.

References

Chapuis, H. and Larzul, C. (2006) How to estimate simultaneously genetic parameters in parental Pekin and Muscovy duck lines using overfed mule ducks performance. *Proceedings of the 8th World Congress on Genetics Applied to Livestock Production.* Belo Horizonte, Minas Gerais, Brazil, 13–18 August 2006, pp. 7–8.

Cheng, Y.S., Rouvier, R., Poivey, J.P. and Tai, C. (1995) Genetic parameters of body weight, egg production and shell quality traits in the Brown Tsaiya laying duck. *Genetics, Selection, Evolution* 27, 459–472.

Cheng, Y.S., Rouvier, R., Hu, Y.H., Tai, J.L. and Tai, C. (2003) Breeding and genetics of waterfowl. *World's Poultry Science Journal* 59, 509–519.

Cheng, Y.S., Rouvier, R., Poivey, J.P., Tai, J.L., Tai, C. and Huang, S.C. (2002) Selection responses for the number of fertile eggs of the Brown Tsaiya (*Anas platyrhynchos*) after a single insemination with pooled Muscovy (*Cairina moschata*) semen. *Genetics, Selection, Evolution* 34, 597–611.

Clayton, G.A. and Powell, J.C. (1979) Growth, food conversion, carcass yields and their heritabilities in ducks (Anas platyrhynchos). *British Poultry Science* 20, 121–127.

Sochoka, A. and Wezyk, S. (1971) Genetic parameters of productivity characters in Pekin duck. *Genetica Polonika* 12, 411–423.

Thiele, H.H. (1995) Recent tendencies in duck breeding. *Misset World Poultry* 11, 31–35.

Thummabood, S. (1992) Recent advances in duck breeds and breeding of Thailand. *Proceedings of the sixth AAAP Animal Science Congress* 2, 83–91.

Xu TieShan, Hou ShuiSAheng, Liu XiaoLin and Huang Wei (2004) Study on synthetic selection index of breast meat weight and breast meat percentage with body size and body weight traits in Peking duck. *China Poultry* 26, 8–10.

Index